commercial forest plantations on saline lands

Marcia Lambert & John Turner

National Library of Australia Cataloguing-in-Publication entry

Lambert, Marcia J.
Commercial forest plantations on saline lands.

Bibliography.
Includes index.
ISBN 0 643 06397 8.

1. Tree farms – Australia – Management
2. Soils, Salts in – Australia.
I. Turner, John, 1947– .
II. Title.

634.9560994

This book is available from:
CSIRO PUBLISHING
PO Box 1139 (150 Oxford Street)
Collingwood VIC 3066
Australia

Tel: (03) 9662 7666 Int: +(613) 9662 7666
Fax: (03) 9662 7555 Int: +(613) 9662 7555
Email: sales@publish.csiro.au
http://www.publish.csiro.au

Printed in Australia by Brown Prior Anderson

FOREWORD

One of the most important factors determining the success of any venture is timing. There could not have been a better time to publish a comprehensive review of the opportunities and the problems associated with the establishment of commercial tree crops or plantations on saline lands.

In an age when environmental issues feature daily in the media it is relatively easy to dismiss yet another environmental problem as an exaggeration or something that will happen perhaps sometime in the future. The salinisation of land and water systems, however, throughout the world and particularly in Australia is occurring at alarming rates and has already caused major losses of potable water supplies and the destruction of many unique ecosystems. For example, in Western Australia, which probably has one of the most accurate documentations of the extent and impact of salinisation of land and water, two million hectares of land in the south-west of the state are salinised, all of the streams originating in agricultural catchments are saline, only three freshwater ecosystems are left. What is even more alarming is the increase in the rate of salinisation. In what are now regarded as conservative projections, it has been estimated that in excess of six million hectares of land will become salinised and 400 species will become extinct within three to four decades in the absence of comprehensive remedial actions. In addition to the impact on the environment, salinisation could have significant implications for our ability to market our agriculture products in the future, as increasingly buyers are demanding that products are certified to meet ecological sustainable criteria.

This book is timely because at last there is widespread community, media and political acknowledgement of the magnitude of the land and water salinisation problem. There is even some recognition that the energies that have been devoted to environmental issues in Australia which are more telegenic but which are infinitely less serious, is one of the reasons why land and water salinisation has received little attention until relatively recent times even though the problem was documented and its cause was correctly identified more than 80 years ago.

Increasingly it is being recognised that the solution to major environmental problems of the scale of land and water salinisation will depend to a significant degree on our ability to collate, coordinate and implement our science with strategies that make profits at the same time as restoring the environment.

The establishment of commercial plantations or tree crops on farms is one strategy that can restore the hydrological balance and result in a profitable return on the investment. But the long term nature of tree crops, the need to plant trees on inherently less productive sites, and the fact that the market for wood fibre, an internationally traded commodity, is intensely competitive means that the costs of growing tree crops on farms must be minimised. Since most tree crops will be established on private land, the cost of land is a significant cost input. Salinised land has no or insignificant value and consequently there is a very large latent low cost land base available for commercial tree crops if the technical problems of establishing trees on saline soils can be solved.

If we are to be successful in addressing Australia's major environmental problem, land and water salinisation, we will need to integrate all our scientific and land management skills. Unfortunately, one of the features of Australian land management and its science is that it is intensely compartmentalised. Land managers are rarely in communication with scientists and often communication and cooperation between scientists is even less than that between scientists and managers. One of the most attractive features of this book is that it provides a synthesis of the science and practical field evaluations of this complex and challenging area. It is also constructed in a way that will enable the scientific and management skills that have been developed to established over one million hectares of plantations in 'traditional' high rainfall fertile sites to be adapted to undertake a planting program ten times that scale on much more challenging sites.

For more than thirty years I have been involved in research, management and the politics surrounding the problem of land and water salinisation in Western Australia and I played a role in achieving the first major incursion of tree crops onto farmland in Western Australia when I held the position of Executive Director of the Department of Conservation and Land Management. I commend this book to scientists, managers, community groups and policy-makers. I believe the authors have been able to synthesise a vast array of information which is an essential prerequisite to an integrated approach to the successful resolution of land and water salinisation. The book will be of international interest not least because there exists in Australia a unique tree gene pool which can be tapped to rehabilitate lands effected by salinisation in many other countries.

Syd Shea

Dr Syd Shea
Principal Adviser
Carbon Sequestration Project
Ministry of the Premier and Cabinet
Western Australia

CONTENTS

Preface

Commercial Forest Plantations on Saline Lands is the outcome of a request by the XYLONOVA Research and Development Syndicate to review available information on saline and waterlogged soils in relation to development of commercial tree plantings. XYLONOVA was established as a multidisciplinary R&D program to bring together the technology, skills and expertise of many of Australia's foremost forest research organisations to develop eucalypt hybrids tolerant of salinity and pollution, together with management practices that maximise the commercial potential of timber plantations. It is anticipated that XYLONOVA will lead to the development of new plantation resources capable of producing high quality timbers suitable for veneer, sawnwood laminated veneer lumber and pulpwood from salt affected lands around the world. Through this focus, the XYLONOVA Program has aimed to treat amelioration of saline lands as a profitable opportunity rather than an environmental cost. Co-sponsorship from Monsanto Company acknowledges a lead role of this corporation in progressing the improvement and productivity of plantation crops, and signals the company's focus on supporting development of environmentally sustainable and profitable solutions to satisfy the increasing global demand for food and fibre products.

During preparation of the book, it became apparent that information on a range of scientific disciplines related to salinity and its potential control needed more collation and analysis than had been originally anticipated. Disciplines including hydrology, soil science, chemistry, tree nutrition, silviculture, genetics, molecular biology, forest products, forest management, catchment management and planning, were all recognised as being important for the success of the project. The objective of the present book is to review relevant information in these areas and provide background for the long-term development and implementation of the XYLONOVA research program. The book is not intended as a manual, but provides the principles that should be considered in establishing and managing plantations under saline site conditions. As such, its message is meant to be universal to owners and managers of degraded lands around the globe.

A conclusion from the review is that the problem of increasing salinity is complex and to address it at any level, input and cooperation from the range of disciples is needed. However, while it was apparent that excellent research is being undertaken in the individual disciplines, there is often a low level of interaction between them, leading to failure to integrate results into practical solutions Further, for an area that attracts so much attention, much of the work undertaken is only available to limited audiences through unpublished literature such as conference proceedings and in-house publications. Such a process limits peer review, distribution of valuable information and practical application of results. It is hoped that the outcomes of this present volume will include greater recognition of the complexity of the problem of salinity and stimulate cooperative work across disciplines, especially in relation to the use of forest plantation trees in saline areas.

About the Authors

Marcia Lambert is a Research Scientist with the forest research company Forsci Pty Ltd working on forest research projects. She was with State Forests of NSW for 38 years, the last ten as Assistant Director of Research and prior to that, Senior Research Scientist in forest nutrition and nutrient cycling. During her career she has been involved in a wide range of cooperative projects with other organisations both in Australia and overseas. She has an Associate Diploma in Applied Science, an Honours Science Degree in Applied Chemistry and an Honours Master of Science in forestry science, specifically on nutrient cycling in forests. She is a member of the Institute of Foresters of Australia, a Fellow of the Royal Australian Chemical Institute, a Member of the Institute of Wood Science, an alumni of the University of Washington, and a full member of the Association of Consulting Foresters of Australia. For ten years, she was the elected representative for forest research organisations in Australia on the International Union of Forest Research Organisations (IUFRO) and in later 1990 was elected to the Executive Board of IUFRO to also represent the countries of the Western Pacific Region. She was honoured to be the first woman to be elected to the Executive Board of IUFRO since its inception in 1892. Current research work is being undertaken in areas such as nutrition and nutritional management of fast grown pine and eucalypt plantations, the carbon dynamics of plantation systems for use in carbon credits, plans of management for Norfolk Island Pines, systems for assessing catchment water quality, regional forestry strategies, commercial plantations on saline lands, etc. She is the author of more than 120 scientific papers and has presented papers at several conferences in Australia and overseas.

Dr John Turner is a Research Scientist currently working with the forest research company Forsci Pty Ltd and also acts as Adjunct Professor in the School of Resource Science and Management at Southern Cross University. He was with the Research Division of State Forests of New South Wales for more than thirty years, initially as a Forester, then Senior Research Scientist and for the last ten years as Director of Research. After obtaining his Bachelors Degree in Forestry Science, he completed a Doctorate at the University of Washington in nutrient cycling in an age sequence of Douglas-fir stands and working with the International Biological Program. Subsequent research work addressed aspects of nutrition, nutrient cycling, and management of forest soils, the focus being on Site Specific

Management. Current research work is being undertaken in areas of nutrition and nutritional management in fast grown plantations, matching appropriate species to site in fast grown plantations, assessment of water quality in catchments, and the carbon dynamics of plantation systems for use in carbon credits. He has more than 150 publications including papers on soil classification, nutritional status, forest/site interactions, fertiliser usage in forests, nutrient cycling, and impacts of forest management.

Acknowledgements

Production of the book was made possible with financial support from the XYLONOVA Research and Development Syndicate and from Monsanto Company and this is very much appreciated. The support of the New South Wales Department of Conservation and Land Management is gratefully acknowledged.

Finalisation of this book involved input, assistance and review from a large number of people and their assistance is very gratefully acknowledged. Our special thanks go to Peter Hopmans, David Flinn, Jim Morris, Jo Sasse, Jen McComb, Nigel Turvey and Glenn Dale. We are also very grateful to authors and institutions for their approval to use published information. Reg Humphreys provided critical comment on parts of the manuscript and we are very grateful for his support. Kyle Lynch assisted with data-basing and analysis of some of the data and her support was also invaluable.

We are sincerely grateful to the following from CSIRO Publishing: Laurie Martinelli for his support of the project; Kevin Jeans and Meredith Lewis who took responsibility for guiding the manuscript to publication; and Brenda Hamilton for her attention to detail in the presentation and layout.

Marcia J. Lambert and John Turner

INTRODUCTION

Plantation opportunities in the saline environment

Soil salinity is increasing in extent throughout the world resulting in large reductions in plant productivity and environmental quality. Planning to both reduce the increase in salinity and to ameliorate areas already affected is complex and expensive. The establishment of forest tree crops on some of the areas will have beneficial environmental effects and also generate economic value. The following discussion considers the magnitude of the problem of salinity and recognises opportunities for commercial and environmental improvement.

Elevated salinity affects approximately one billion hectares of land surface and about one-third of all irrigated land, and is considered to be a significant factor in reducing agricultural production (Maas and Hoffman 1977; Hasegawa *et al.* 1986). Such problems are not recent. Disastrous long-term effects of salinity have been described for a number of ancient civilisations (Jacobsen and Adams 1958) but the extent of the problem is now on a much larger scale. Various processes are involved in causing salt to increase and consequently, there is a range of differing conditions. However common to all situations is a cycle of elevated levels of salts caused by land use changes, resulting in reduced biological production which leads to further increases in salt accumulation. One method of breaking this cycle is with the use of tree species to increase biological production, preferably using a commercial tree crop. This results in beneficial environmental effects and the high commercial value assists with amelioration of the social and economic problems resulting from increases in salinity.

Soil or water salinisation is the process by which salt accumulates in the soil or water to such an extent that it adversely affects the use of the soil for plant growth, or the water for supply to humans, stock or industry (Williamson 1990). Saline soils contain levels of soluble salts high enough to reduce or eliminate plant growth. There are many areas where the salinity levels in naturally saline water or soils have always restricted the range of possible uses. This naturally high level is called primary salinity and includes salt marshes, salt flats and salt lakes, some in coastal regions and others inland in semi-arid or arid regions. All such areas are associated with highly saline ground waters and often with impeded internal drainage. While these areas provide valuable information on salinity, those which have become saline due to human activities are the present main focus. Basically, increased salinity from man-made causes develops when natural deep rooted vegetation is removed and replaced with shallow rooted crops and pasture, particularly in areas with limited rainfall. In addition, irrigation can lead to rising watertables and salt accumulation. Such salinity is termed secondary salinity. The processes of change both in vegetation and irrigation add to the complexity of the overall problem.

While salinity is often simply defined as high levels of salt, it is in fact a complex biological, environmental, commercial and social problem. A major component of salinity relates to sodium chloride, either through its effects on plant-water relations or the toxicity of the sodium and/or chloride ion. There are also interactions with a range of other ions. Such interactions may further complicate the problems of water salinisation due to the presence of sulfates and/or carbonates, or can result in other nutrient toxicities or deficiencies such as those associated with calcium, cadmium, zinc, aluminium or manganese. The optimisation of ameliorative procedures requires definition of the specific problem on each site and the development of an appropriate strategy. Any ameliorative procedure is expensive and this is also true of tree plantations. The costs involved in the establishment of tree plantings need to be covered by governments or landowners by establishing sufficiently large areas of productive commercial plantations to yield financial returns to cover the amelioration activities and provide incentives for long-term management. The identification of markets for the products is a critical primary component of such developments. Major characteristics of salinity are the high variability, the very large areas affected, and changes over time. This means that the management solutions need to be innovative and applicable on a broad scale. The approach of tree planting is not one of developing plantations on the same site in perpetuity. It is a process of maintaining planted trees in the landscape on a cyclic basis as an aid to ensuring that the overall productive capacity of land is increased and maintained.

There are many areas where trees will not be suitable, but there are large areas where trees will be a significant component of the ameliorative process. This book provides information on the principles and practices to establish plantations at a scale sufficient for economic value and provide regional environmental benefits. This involves policy development, regional and farm level planning together with operational management. Implications for research are also considered.

General effects of increasing salinity

Increases in salt produce impacts at the small and large scale. Soil salinity not only reduces plant productivity but also detrimentally affects the quality of run-off water. The effects of increases in salinity on a broad scale can be shown from an example in Australia. Elevated salt levels in soils are affecting large areas of the Murray–Darling Basin located in the states of Queensland, New South Wales, Victoria and South Australia. Broadly, salinity in Australia is higher in the southern areas of the country and is related to winter rainfall and broadscale regions with geologic restriction to ground water flow (Shaw 1999). It was estimated by the Murray–Darling Basin Ministerial Council (1989) that in 1987 the salt loads from dryland areas would eventually result in a rise in electrical conductivity of between 30 and 50 μS/cm in the River Murray at Morgan in South Australia. Recent studies have indicated that this estimate was too low. The prediction has been made that within the Murray–Darling Drainage Division alone, the area affected by dryland salinisation will increase by the year 2010 from 2000 square kilometres to 10 000 square kilometres. It has been estimated that even with beneficial management regimes in areas adjacent to the Murray, there will be an increase of 37 μS/cm in run-off water salinity and that continued clearing of the Victorian Riverine Plains could produce an increase of 140 μS/cm (Williamson *et al.* 1997).

The Salinity Audit (1999) enlarged and refined existing knowledge of the threat of salinity and established a trend, river valley by river valley, for salt mobilisation in the landscape and its expression in rivers and at the land surface. A very important finding was the estimation that much of the salt mobilised does not get exported through the rivers to the sea. It stays in the landscape or gets diverted into the irrigation area and floodplain wetlands. The Audit identified the severity and scale of the salinity threat to the Murray–Darling Basin if there are no new management interventions.

The economic and social constraints imposed by land degradation are substantial. The cost of losses to agricultural production in the Murray–Darling Basin as a result of salinity were estimated by the Murray–Darling Basin Ministerial Council (1989) to be $260 million per annum with further losses of $69 million due to deterioration of urban water supplies. In Western Australia, losses in productivity due to salinity have been valued at approximately $200 million per annum. With such high losses it may be considered that investment in ameliorative measures could be readily justified. However estimates were made on the use of trees for amelioration and the costs of tree establishment were so high that tree planting could not be considered as a strategy for ground water and salinity management (Dumsday *et al.* 1989; Dumsday and Oram 1990). However these costs were partly structural (tax and marketing) rather than biological, while trees and new genotypes are practical options for the future. It is concluded that further evaluations of the economies of large-scale tree planting with direct and indirect benefits need to be undertaken. There are very limited examples in Australia and while it is difficult to provide comparable cost figures on a regional or world basis, the impacts are very large and increasing. A primary focus has been on crop losses and the effects on water resources, and while they stress the variability of the problem, none attempt to reduce the importance of the problem and its impacts (Ghassemi *et al.* 1995).

In Australia, much of the ameliorative work being undertaken is in relation to land care and total catchment management programs where the primary focus is to improve soil and water conditions. Commercial forestry crops are currently of secondary importance. The planting of trees in the programs is clearly an important aspect of amelioration but information on suitable species and establishment conditions is not well developed. Additional requirements to develop tree plantations on a commercial as well as an ameliorative basis are even less well understood, particularly in relation to maintaining a minimum level of productivity over a large area. The establishment of plantations in typical rainfed areas is well understood but where there are significant growth limiting factors, such as salt, and especially where they change over time, they are poorly understood.

Principles associated with salinity for Australian conditions have been previously emphasised and they also apply to other areas:

- Salinity is a natural component of the Australian landscape.
- The community and the individual will only invest in salinity control if the benefits are significantly greater than the costs.
- An economic incentive in the form of profitable land use or community support is an essential component of any program.
- Land and water have finite values and their value will govern what the community or farmer is willing and able to pay for research and management.

These issues need to be kept in mind when developing projects in relation to salinity and a major consideration is how to derive economic benefits to overcome the environmental problems.

Research and Development

Research and Development programs are key issues in long-term salinity amelioration. The suitable establishment of plantation trees in saline areas will require a strong research effort at all levels from planning and policy, through landscape issues to specific biological aspects. It is apparent that there is little research integrating the biological, soils, hydrological, and management aspects of forest plantations and unless overcome, this will be a problem in broader scale developments.

One recent focus has been on genotype selection. Selection of tree species for planting in areas with elevated salinity has been made on the basis of their salt tolerance and survival in saline conditions. Hence the measure of planting success has been the survival of individuals, rather than the productive capacity of particular species. The commercial value of such plantings, particularly the establishment of sufficient numbers of trees to allow for marketing to be undertaken and hence sustain a commercial enterprise, has been of secondary importance.

A large number of coordinated projects will be required to minimise both the expansion of salinity and to reduce the areas currently affected. They will need to be at regional, catchment and individual property levels, and will need to be an integration of engineering, land management practices, and biological approaches. One component of biological amelioration is the use of deep-rooted perennials such as trees with suitably high productivity to utilise soil water and hence lower the watertable and salinity. Biologically, there needs to be the development of genotypes which have amelioration benefits, high productivity and desirable end products. Research on genetic development and molecular biology is required to support this approach and needs to include silvicultural practices to ensure optimum management (fertiliser and pruning treatments; and weed, pest and disease control strategies), and wood science and product development. Much of this research is still at a basic level, for example establishing a standard methodology for assessing salt tolerance in trees.

The key aims of such research programs would include:

- To identify a commercial market for products.
- To define and identify site characteristics and associated problems.
- To produce suitably productive genetic planting stock capable of growing in specifically identified saline situations.
- To have well defined plantation management prescriptions.
- To develop systems to evaluate the commercial and environmental benefits of the planting system.

Further, research needs to be undertaken at various scales to determine impacts of scales of planting and where trees are best located in the landscape at any time. Aspects of these areas are discussed in the book, recognising the technical support for such plantations will be a key issue.

Features of this book

The book focuses on published information which is currently available concerning the establishment and management of commercial forest tree plantations in difficult environments under sub-optimal growth conditions, which often occur in conjunction with waterlogged soils and are caused by high levels of salt in soils or applied water. The conditions may have been brought about by natural circumstances or induced by management practices such as irrigation. The critical factor is that trees with appropriate timber characteristics or other properties can be grown on a sufficiently large scale to maintain a commercial operation. The properties, physiological characteristics, scale of operation and management practices are addressed for such situations.

The primary species of interest are in the genus *Eucalyptus* because there are identified potential utilisation characteristics and markets. Other commercial species, specifically exotic conifers, are referred to. Tree species in genera such as *Acacia*, *Melaleuca* and *Casuarina* are recognised as having high salt tolerance and have been studied in relation to fuelwood use, but are not a focus of this review because they are not a top priority for production of industrial timber.

Issues involved in defining the characteristics of sites where plantations may be established and their special management requirements are discussed. Central to the development of such projects is the selection and development of appropriate genotypes and appropriate management practices, and options for this are presented. Finally, since rapid changes occur over time within plantations developed on saline and waterlogged areas, the monitoring of the plantations is shown to be a vital management issue. Essentially commercial tree plantations for amelioration of saline environments are a viable option requiring careful scientific management.

Basically the best application of forest plantation practices with appropriate genotypes can lead to environmental benefits and commercial opportunity. The book aims to enhance these opportunities.

CHAPTER 1

SALT IN THE ENVIRONMENT

Issues

Clearing of native perennial vegetation and other land use changes result in water, and the contained salt, moving toward the soil surface. This is especially the case in lower rainfall areas. Such effects are extensive and both temporally and spatially variable in their characteristics, but ultimately they detrimentally affect plant productivity. One component of undertaking effective ameliorative measures is the characterisation of the saline environment. Systems for doing so are presented.

Salinity in the landscape

Salt-affected soils cover more than 7% of the world's land area in that approximately one billion hectares are detrimentally affected by high levels of salt in the soil (Szabolcs 1981; Dudal and Purnell 1986). A large proportion of this area is primary saltland (naturally occurring) such as coastal saline marshes, mangroves, estuarine and coastal plain floodlands, and internally drained lakes. However, an increasing proportion of the land surface is becoming saline due to secondary causes, particularly as a result of agricultural clearing and irrigation. Secondary saline areas are often segregated into salinity associated with irrigation and that with non-irrigated or dryland areas.

Dryland salinity

Dryland salinity results from clearing the original deep-rooted perennial vegetation cover from areas with salt stored in the soil profile or those with saline ground water, and replacing it with shallow rooted annual crops and pastures (see Box 1.1). Reduced evapotranspiration due to the removal of the native vegetation results in a changed hydrological cycle, primarily increased ground water recharge and leaching of salt into ground waters, causing watertables to rise and become saline (Burch *et al.* 1987). Over time, ground water has risen into the root zone of plants bringing with it dissolved salts from further down the soil profile. Simply, excessive clearing of native vegetation on hill slopes has increased the amount of rainfall re-charging ground water systems, thus causing watertable levels to rise. The rising water levels cause increased mobilisation of soil salts and subsurface seepage flow across the landscape, and saline seepage emerges lower downslope in situations determined by landform and surface geomorphology. The effects of seepage result in declines in growth and eventual death of plants on the site, and the development of bare areas which yield high rates of run-off and may lead to the initiation of gully erosion (Dye 1979; Charman and Junor 1989). The period over which effects appear after clearing can be variable, with saline seeps often taking 50 to 60 years to appear (Wagner 1989). The rising salts can become concentrated in the surface by evaporation of water, causing scalds. In other cases the ground water stays at the soil surface, causing seeps.

Estimates of areas affected by dryland salinity on a global basis vary greatly but the areas are very significant. One estimate is about 76.6 million hectares globally (Table 1.1) although other workers have reported much higher estimates of saline and sodic soils (Abrol *et al.* 1988). Areas affected are difficult to estimate as they change rapidly and are caused by factors such as the rainfall in any one year (Charman and Junor 1989). Extended periods of above average rainfall followed by extended dry periods increase the extent of the salinity area and the salt problems. The estimate of 0.9 million ha of dryland salinity for Australasia (Table 1.1) is lower than the estimate of 2.2 million ha by McLennan (1996) and the 3.8 million ha of affected land reported by Poulter and Chaffer (1991) and Williamson (1990), but higher than the 0.8 million ha estimated by Schofield (1992). In most states this area is increasing, with the rate of increase reported as 5% per annum (Pepper and Craig 1986). The estimate of about 0.4 million ha of extremely salt-affected land (Table 1.1) is relatively consistently reported, however there is variability occurring in the other categories primarily indicating the high variation and uncertainty associated with estimates of salt-affected areas.

Dryland salinity is a major problem in south-west Western Australia, southern South Australia and northern Victoria. In Western Australia, 443 000 ha or 28% of cleared land has become salinised and the annual rate of salinisation is approximately 18 000 ha/yr (Schofield

Box 1.1 Dryland salinity development resulting from watertable rise (Anon 1989).

Dryland salinity is the build-up of salts in the surface soil, usually as a result of a rising watertable and subsequent ground water seepage. These salts have been stored to depths of 30 m or more in many soil types and have accumulated there from several sources: the ocean via rainfall, weathering of soil and rock minerals, and marine deposition in earlier geological periods. Watertables have slowly risen in response to clearing of deep-rooted native vegetation, including trees (see figure below). The clearing has allowed large volumes of rainfall to leak through the soil and enter ('recharge') the ground water system. This water re-surfaces in 'discharge' areas.

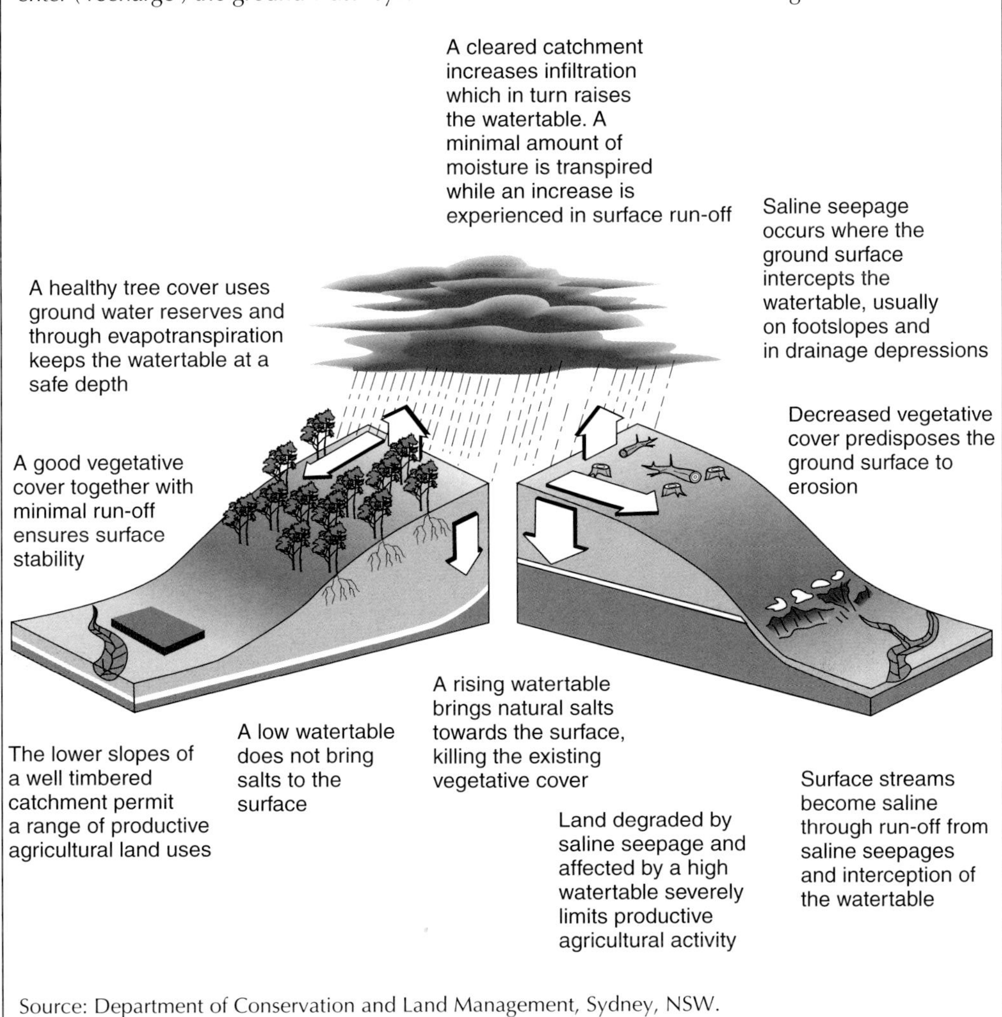

Source: Department of Conservation and Land Management, Sydney, NSW.

1989; Schofield and Bari 1991). In addition, 156 000 hectares are considered to be affected by irrigation salinity in Australia; this problem relating to the increased build-up of salts in the soil and water systems. It is caused by rising saline watertables and the repeated use of salinised river water for irrigation (Charman and Murphy 1991).

Table 1.1 Regional estimates of dryland salinity (human-induced) (Oldeman *et al.* 1991; Ghassemi *et al.* 1995).

	Dryland salinity (million ha)				
Region	**Light**	**Moderate**	**Strong**	**Extreme**	**Total**
Africa	4.7	7.7	2.4	–	14.8
Asia	26.8	8.5	17.0	0.4	52.7
South America	1.8	0.3	–	–	2.1
North & Central America	0.3	1.5	0.5	–	2.3
Europe	1.0	2.3	0.5	–	3.8
Australasia	–	0.5	–	0.4	0.9
Total	34.6	20.8	20.4	0.8	76.6

Secondary salinity associated with irrigation

Secondary salinity induced by irrigation is found in many countries which have arid and semi-arid climates. Reported estimates of such salt-affected lands on a global basis are in excess of 28 million hectares but considering the lack of reporting by some countries, the true estimate is very much higher (Table 1.2). The area is approximately half the area affected by dryland salinity but the productive capacity of these lands is much greater. Ghassemi *et al.* (1995) estimated a smaller area of 5.4 million ha (20%) of the world's irrigated lands are salt affected but much larger areas are at imminent risk due to shallow watertables. Again, the estimates are variable indicating the problem of actually determining the extent of such a serious problem.

Table 1.2 Estimates of secondary salinity (salt-affected and/or waterlogged irrigated lands) (Chaudary *et al.* 1978; Zohar 1979; Goldsmith and Hildyard 1988; Ghassemi *et al.* 1995; Gafni 1997).

Country	**Area affected (million ha)**
India	7.00
China	6.70
Pakistan	4.22
USA	4.16
Iraq	1.80
Egypt	1.70
Argentina	1.58
Australia	0.77
Peru	0.30
Israel	0.11
Total	28.34

Table 1.3 Extent of current human-induced salinity of soils in states of Australia (Williamson 1990) ordered from left to right according to decreasing area of scalds in secondary salinised soils.

	Area affected ('000 ha)							
	SA	NSW	NT	Qld	WA	Vic.	Tas.	Total
Secondary salinised soils								
Scalds	1200	920	680	580	340	60	0	3780
Saline seeps (non-irrigated)	225	14	0	8	443	100	8	798
Irrigated saline soil	0.5	10	0	1	0.5	144	0	156
Shallow ground water (<2 m) in irrigated lands	4.5	260	0	0.5	0	385	0	650

Secondary salinity which has ground water as a key component results in impacts on soil, vegetation and water resources. The salinity develops due to the discharge of ground water. In irrigated areas, these saline soils are frequently termed scalds (although the term 'scald' is also applied to exposed laterally saline subsoil as in Table 1.3), but in non-irrigated lands, the salt affected areas are termed saline seeps. A saline seep is an area where salts accumulate at the soil surface as a consequence of ground water discharge from a confined or unconfined aquifer. Areas of secondary salinity within Australia have been estimated according to the type of effect (Table 1.3).

Secondary salinisation in both irrigated and dryland situations, and problems arising from surface and subsoil sodicity, have counterparts throughout the world but the emphases, especially in the Northern Hemisphere, appear to be different to those in Australia (House of Representatives, Standing Committee on Environment, Recreation and the Arts 1989). In the case of irrigated soils, this is because much of Australia's irrigation water is applied to relatively impermeable clays and duplex (texture contrast) soils rather than the more permeable silty and sandy soils commonly cultivated in India, China, Central Asia, Eastern Europe and the USA. In the dryland context, the widespread use of sodic duplex soils for farming operations in Australia has led to hillside seepage, valley salting, tunnel erosion, and scalding. Such degradation phenomena have received scant attention in the world literature possibly because they are of less significance elsewhere. Robertson (1996) attempted to estimate rates of change in salt seep areas, suggesting that there are problems with field estimates (Table 1.4) due to temporal variation and changes in intensity.

In all the areas with secondary salinity, establishment of woody vegetation, shrubs and trees assists with amelioration. Where water availability is higher due to precipitation, ground water or irrigation, commercial crops of plantation trees can be grown leading to amelioration and commercial returns.

A secondary process of alkalinisation is associated with soil salinisation where the soil exchange complex is saturated with sodium. Sodium ions lead to dispersion of the fine clay particles and cause the soil structure to collapse. This leads to less permeable soils for water infiltration and aeration (Hillel 1990).

Table 1.4 Areas (ha) of salt seeps in 1982 and in the period from 1991 to 1995.

	1982	Latest available data in the period from 1991 to 1995	Annual change over approx 11 years
New South Wales	0	25 000	2 300
Victoria	90 000	120 000	2 700
Queensland	0	10 000	1 000
Western Australia	264 000	1 609 000	123 000
South Australia	55 000	401 000	31 000
Tasmania	0	N/A	-
Northern Territory	0	0	-
Total	409 000	2 165 000	160 000

Land management change and salinity

Salinity can change and be highly variable with respect to the composition of salt involved. For example, in Australia, Williamson (1990) emphasised a high level of sodium and chloride, whereas in India and other countries there are significant areas with high sulfates of calcium and magnesium and high pH (Mathur and Sharma 1984). Similarly, work in the USA has shown sulfate to be the dominant anion with excessive concentrations of boron (Shannon *et al.* 1998). Such variation in salt composition affects the basic mechanisms involved and also affects plant species tolerance.

On a global basis, there is considerable variation in the climate where salinity occurs. Areas with salt-affected soils generally have semi-arid and arid climates (for example Australia and northern India), but include seasonally dry tropical climates such as in Thailand (Arunin 1984, 1987)), semi-arid coastal climates (for example in India) and sub-humid and humid climates (for example south-west Western Australia and coastal Queensland). The variation in climate will affect the variability of salinity both seasonally and in the longer term. In India, the problem of soil salinity/sodicity is considered one of the most adverse factors restricting economic utilisation of available land resources. It is estimated that about 7 million ha are affected in India, primarily in the Indo-Gangetic plains, arid and semi-arid areas of Rajasthan and Gujaret, areas of black soils and on coastal tracts. The areas affected by soil salinity are increasing mainly due to the expansion of irrigation projects coupled with poor water management, often leading to waterlogging (Yadav and Singh 1970).

The management of salt-affected soils is of considerable significance in Australia because of the wide and extensive distribution of these soils throughout the continent. It has been estimated that saline (primary and secondary) and sodic soils together cover about one-third (2.4×10^6 km^2) of the total area of Australia (Northcote and Skene 1972). Saline soils occur mainly in the more arid interior, while sodic soils, many of which have saline subsoils, are especially common in south-western and south-eastern Australia. For climatic and topographic reasons, these are areas in which a significant part of the nation's mixed farming, involving cereal cropping and grazing of native and improved pastures and irrigation, is practised. Considerable disturbance of the natural ecosystems, such as clearing of native vegetation has usually occurred, with consequent changes to hydrologic regimes. Salt formerly

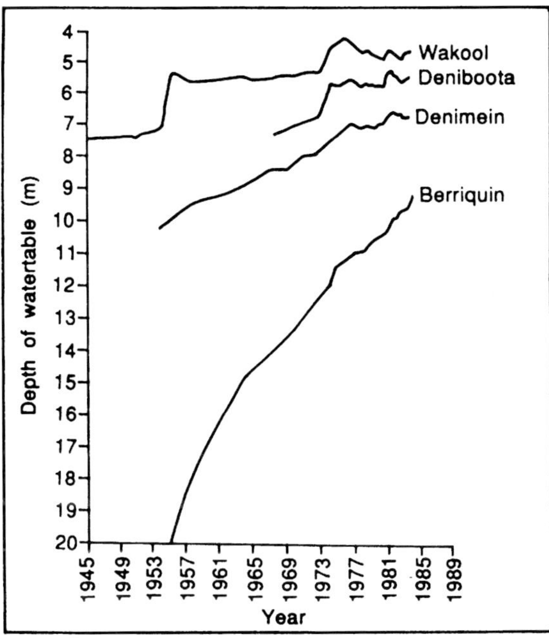

Figure 1.1 Changes in depth to watertable within irrigated areas of the Murray–Darling Basin, Australia (Gutteridge *et al.* 1970, Murray–Darling Basin Ministerial Council 1989; Ghassemi *et al.* 1995).

distributed in deeper subsoil and substrate layers in the catchments and alluvial plains has become mobilised and concentrated in the vulnerable parts of the landscape by the process of secondary salinisation (Loveday and Bridge 1983; Dudal and Purnell 1986).

The relationship between land clearing and salinity in Australia has been known for more than 100 years (Robertson 1993). For example, in the late 19th century as part of the Goldfields Water Supply Scheme in Western Australia, the catchment of the Mundaring Weir was partially cleared to increase water supply to the dam. Within a few years, the water retained by the weir showed increases in salinity. In 1907, the increased salinity was associated with land clearing. Subsequently the catchment was re-vegetated by regrowth and forest plantations and the water salinity levels declined. In relating this example, Robertson argued that in 100 years research has provided little more information than was available then. The assessment of salinity has been quite subjective, with the expression of the problem as detected by transects, aerial photography or remote sensing being the main means of defining saline land (Robertson 1996). Neither the actual or potential increase in salt content is usually specified, nor its impact on plant growth. Hence each method has provided variable results depending on the climate, seasonal conditions and criteria established by the assessor.

In addition to salinity increasing, there is the problem of rising watertables leading to waterlogged soils. In the Murray–Darling Basin in Australia, poor irrigation practices in the irrigated areas which currently cover more than 1.2 million ha, have led to a rise in watertables of between 10 and 20 cm per year since the last half of the 19th century (Ghassemi *et al.* 1995). The watertable rises have been monitored (Fig. 1.1) and there is little evidence of

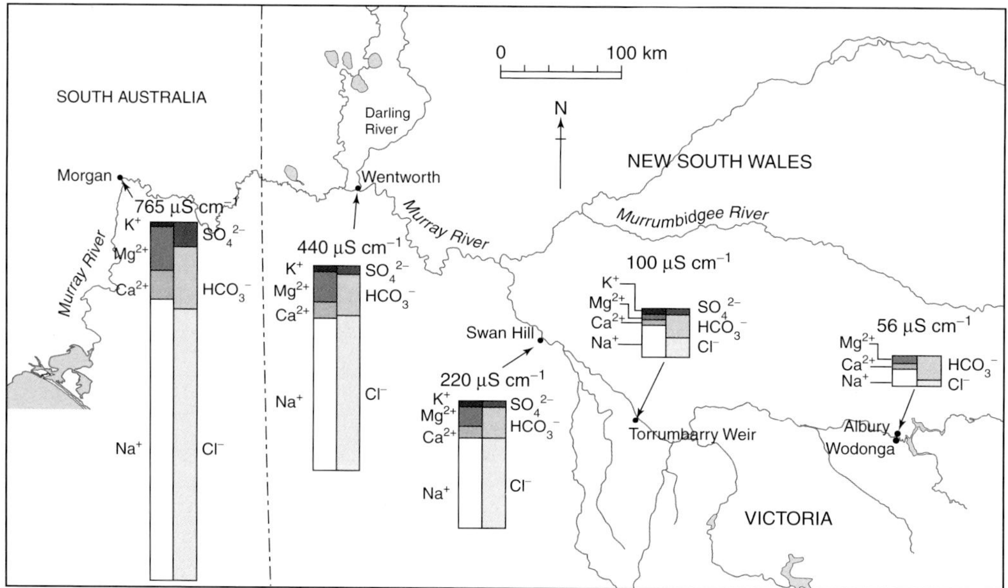

Figure 1.2 Variation in electrical conductivity and proportions of the major cations and anions in river water according to distance down the Murray River. Values are based on median concentrations for 1979–80 with ionic proportions calculated using molar ratios (from Hart and McKelvie 1986).

equilibrium being reached. In addition, the water is increasing in salinity as it moves down the river (Fig. 1.2).

Changes associated with soil degradation will affect the long-term sustainability of activities dependent on these soils. Costs arising from soil degradation have effects both on-site and off-site and, while they are difficult to quantify, are obviously high. The costs of salinity to agricultural production alone in the Murray–Darling Basin have been estimated as $A1660 m per year (Pearce 1999). When extended to other areas, the effects on agricultural production costs are enormous.

Processes related to soil salinity

For salt to become an environmental problem, there are three basic requirements:

- source of salt
- source of water in which the salt is dissolved
- mechanism by which the salt is redistributed to locations in the landscape where it can be damaging, including into rivers (Williamson 1990).

Sources of salt

Salt may be derived from a number of sources which include salt derived from weathering of original soil parent materials, cyclic salt deposits from rainfall, salts deposited from capillary action from shallow ground water tables, salt seepage from upland areas, and salts contained in applied irrigation water. The importance of each varies regionally. By world standards, most Australian agricultural soils are relatively old in geological terms, with deeply weathered or thick alluvial deposits of clays under the cultivated zone. Many Australian soils contain high quantities of salt stored at depths of 30 m or more. Soil surveys have shown that about 238 million hectares have high quantities of salt and about half of these have been intensively managed for agriculture. However, surveys usually only identify the salt content of the soil layer which is usually considered to extend about 1 m below the soil surface. The total salt content for the whole soil profile to basement rock has been studied extensively in the south-west of the Australian continent. There is a systematic increase in stored salt with decreasing rainfall, ranging from 200 000 kg/ha at about 1000 mm/year rainfall to 1 million kg/ha at 600 mm/yr rainfall. Slightly lower quantities of salt have been measured in the northern slopes of Victoria over a similar rainfall range. Approximately 75% of this salt is stored in that part of the profile which was unsaturated with water at the time of vegetation removal. For the deep sandy soil profiles of the Murray Mallee in South Australia, the unsaturated zone is quite low in stored salt to depths of about 30 m compared with the ground water zone beneath in which the greater part of the salt exists.

Salt which has accumulated in the soil may originate from several sources: the ocean via current rainfall, weathering of soil and rock minerals, and marine deposition in earlier geological periods. Input of salt via rainfall has been measured in Australia at 300 kg/ha/yr near the coast, about 30 kg/ha/yr at 250 km inland, and about 15 kg/ha/yr at greater than 600 km inland. The total salt input to the 106 million hectares of the Murray–Darling Basin from rainfall in 1974–75 was measured and found to be about one million tonnes per year (Blackburn and McLeod 1983). The rate of release of salt due to weathering of soil and rock minerals varies greatly, but additions to the soil profile are considered to be less than 1% of that brought in by rainfall. Using the present data on salt inputs, the measured levels of salt storage could have been accumulated during the last 60 000 years (Williamson 1990). In irrigation areas, the salt stored in the soil is further increased by that in the applied irrigation water. The application of water to a depth of 1 m with a salt concentration of 500 mg total soluble salts (TSS)/L adds 5000 kg of salt to each hectare of irrigated land. These data add to the estimates of accumulations which have been the major sources of salt in recent geological times. The salt is located in both the unsaturated zone of the profile and the accumulated ground water where concentrations as high as 20 000 mg TSS/L are not uncommon (Williamson 1990).

Generally, deposition is considered to be a major source of salt (Peck 1977; Turner *et al.* 1996a) and in undisturbed ecosystems there is an equilibrium in that salt inputs and outputs are equal, but in agricultural systems in higher rainfall areas outputs are very high (Peck 1977). In higher rainfall areas generally salt is readily leached through the soil, but where rainfall is lower it will accumulate. Chloride and sodium inputs were related to distance from the coast in eastern Australia: that is, there are very high inputs near the coast, rapidly declining further inland (Turner *et al.* 1996a). This pattern is fairly consistent in other areas, for example in Western Australia (Fig. 1.3).

Sodium and chloride have been the two main elements of interest in salinity issues, and inputs of these elements vary greatly on an annual basis. Typical values of inputs are given in Table 1.5 with the highest levels generally close to the coast; for both elements, the range was from 2 to 400 kg/ha/yr for a range of sites. Although Bettenay *et al.* (1964) concluded that most of the salt in the landscape of a catchment in Western Australia was from long-term accumulation of atmospheric inputs, the effects of salt inputs on vegetation such as forest plantations in salt-affected areas are considered to be minimal. Studies on small

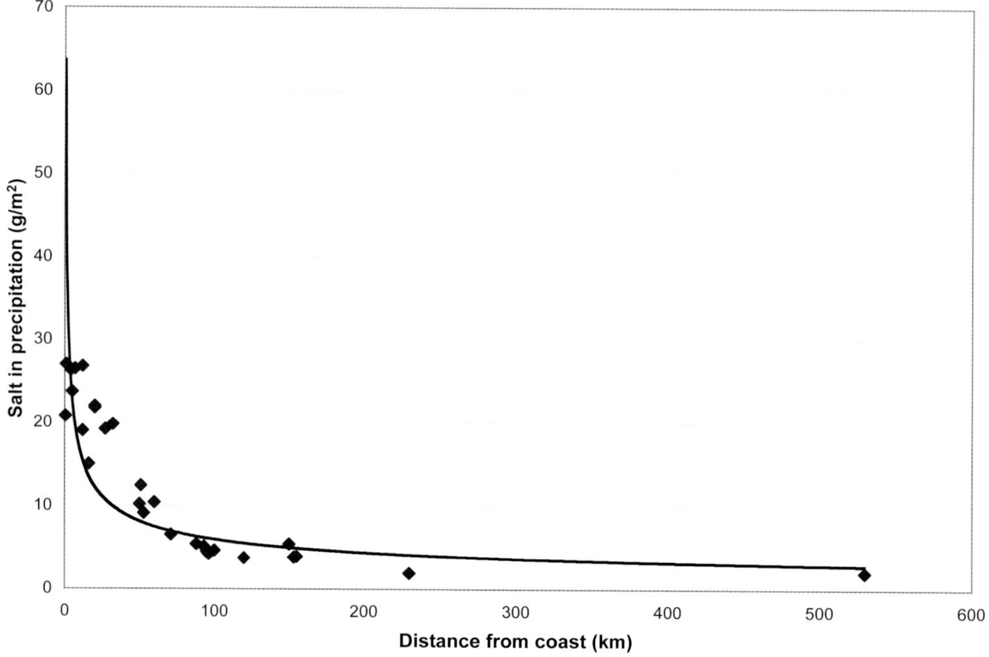

Figure 1.3 The quantity of salt precipitated by rainfall and dry fallout and the effective salt concentration in rainwater in southwest Western Australia in relation to distance from the coast (Borg *et al.* 1987; Hingston and Gailitis 1976).

Table 1.5 Annual inputs of sodium and chloride at selected locations in New South Wales (Turner *et al.* 1996a).

Location	Distance from coast (km)	Sodium input (kg/ha/yr)	Chloride input (kg/ha/yr)
Coffs Harbour	1	23	42
Karuah	15	15	24
Eden (research weirs)	30	14	25
Newnes	114	7	8
Armidale	135	3	2
Narrandera	350	7	11
Baradine	366	2	4

catchments showed that as a result of partial clearing, catchments changed from net chloride accumulators to net chloride exporters (Ruprecht and Schofield 1989, 1991a, 1991b). The losses of chloride associated with clearing were demonstrated over a period of time in Western Australia (Fig. 1.4). In a similar area, Sharma *et al.* (1987) demonstrated that within two years of clearing, there were highly significant increases in soil water storage. The conclusion is that at the smaller scale, there is going to be high annual variation in effects and that constant monitoring is required to understand patterns.

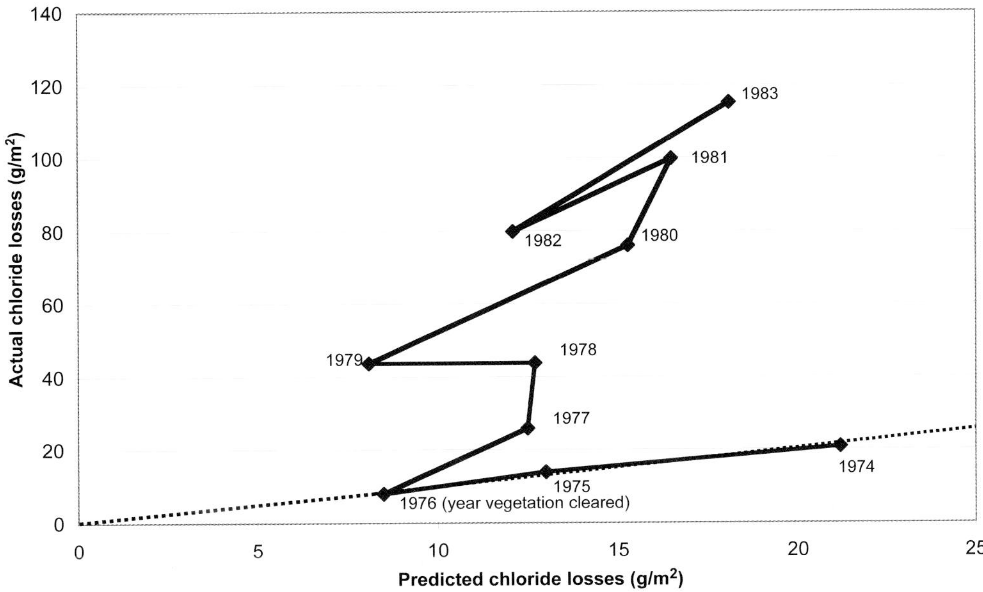

Figure 1.4 Predicted and actual chloride losses from a forested then cleared catchment in Western Australia. The predicted values are modelled from a coastal (untreated) catchment which was cleared in 1976. The actual losses (g/m^2) are continuing to increase (after Macpherson and Peck 1987; Williamson et al. 1987).

Table 1.6 Relationship between spatial scale and response time in saline environments.

Field of view	Spatial unit	Response time
km^2 x 1000	Large catchments	Decades
km^2	Small catchments	Years
m^2	Seeps/Catena	Months/Years
m	Soil water Vertical change	Years
mm/cm	Rhizosphere/Colloids	Months

The spatial scale of salinity studies has been highly variable and as such, response times and degree of variation are affected. In large catchments considered as thousands of hectares, responses to disturbance (clearing or reforestation) will be in the order of decades (Table 1.6) whereas seeps within the landscape are usually measured in square metres and will respond to change in months or years. The long time periods at the landscape level and the high variability require the development of sound models for suitable analyses of responses

Table 1.7 Inputs and outputs of sodium and chloride in a small forested catchment (Flinn *et al.* 1979; Hopmans *et al.* 1990).

	Inputs	Outputs	Difference
		(Equivalents/ha)	
Sodium	0.112	0.116	–0.004
Chloride	0.386	0.096	+0.29
Excess over equivalent balance (Na–Cl)	–0.274	+0.020	
% Excess of chloride	71%	21%	

(Loh and Stokes 1981; Hookey 1987; Macpherson and Peck 1987; Allison *et al.* 1990; Tanji and Karajeh 1993; Hatton *et al.* 1993; Pierce *et al.* 1993; Walker *et al.* 1993; Karajeh and Tanji 1994a, 1994b).

Sodium and chloride inputs are not necessarily chemically balanced, often chloride in higher rainfall areas largely balances the total of the cations. The sodium and chloride inputs and outputs on a catchment basis may not balance in any given year, as sometimes there is an apparent accumulation of sodium and at other times, chloride (Table 1.7).

Large accumulation of salts in soils can be accounted for by inputs from precipitation over recent geological history. The salt inputs in precipitation are primarily sourced from oceans and measured inputs are higher closer to coasts. However, except near the coast, inputs of salt in rainfall do not appear to have significant effects on vegetation. Critically, inputs and outputs are very variable on an annual basis and patterns of change may be slow.

The source of water

The replacement of deep-rooted perennial plants with shallow rooted annual crops and pastures produces a reduction in evapotranspiration and an increase in net precipitation reaching the soil surface. Removal of the native vegetation decreases the interception loss by between 9 and 13% of rainfall. The overall result is an increase in the volume of water draining below the root zone in agricultural areas, and a cessation of any withdrawal of water by plants from the ground water during periods of low rainfall. Consequently ground water levels rise and there will be an increase in the volume of water moving towards the discharge sites, either at a seepage area or directly into a stream at a lower level in the landscape.

The initial effect of land clearing is usually an increase in the quantity of ground water moving into the stream so that the salinity impact is dependent on the salt content of the ground water nearest the stream (Fig. 1.5). The clearing of land also increases the amount of surface run-off which, because of its low salt content, dilutes the additional salt added to the stream from ground water. This was demonstrated where land clearing greatly increased quantities of chloride in run-off water at one site in Western Australia (Fig. 1.6). The chloride was derived from soil storage.

Mechanisms for redistributing salt

Mechanisms for redistributing salt in the landscape are complex but a number of hydraulic factors are important. Ground water conveying salt to the seepage zones has to move

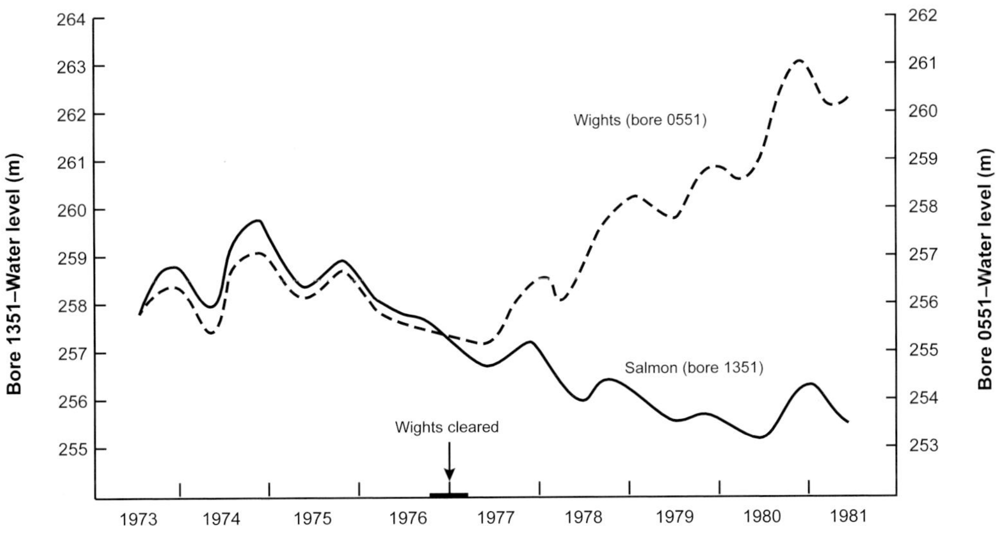

Figure 1.5 Effects of land clearing on the watertable in two Western Australian catchments: the Wights catchment was cleared in 1976–77, while the Salmon catchment remained forested (Peck 1983).

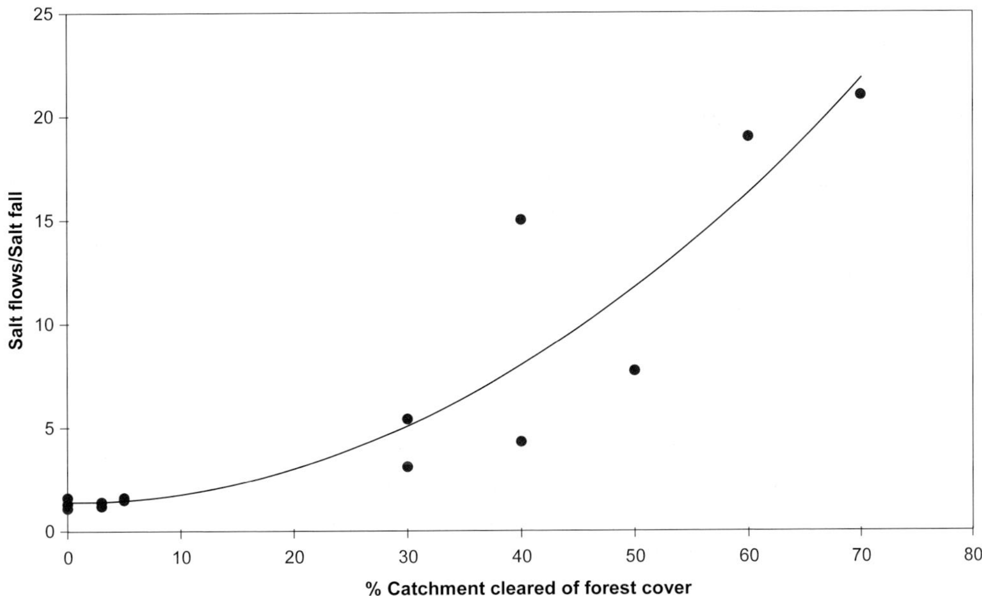

Figure 1.6 Ratio of salt in run-off water to input in rainfall in relation to the proportion of forest area cleared in a series of catchments in Western Australia (Peck and Hurle 1973).

through weathered rock material such as clay which has a considerable resistance to the movement of the water. Frequently there is a two-layered ground water system. The shallow aquifer is usually the dominant source of water while the deep aquifer is the major source of salt, discharging into saline seeps and streams.

Factors influencing the mobilisation and redistribution of salt in the landscape reflect the complexity of ground water systems and identify the potential influences which need to be considered when developing a conceptual model of salt mobilisation. The relative importance of any particular factor varies from catchment to catchment and, in some cases, may be of limited or of no significance and include:

- Climate, rainfall and salt fall (from precipitation) distribution
- Ground water hydrology including hydraulic head trend, hydraulic head gradient and hydraulic conductivity
- Ground water chemistry including ionic ratios, pH and aquifer salt storage
- Land use, both present and historic, length of time since the establishment of agriculture, ecological systems and land management systems
- Surface water hydrology and chemistry, wetlands, lakes and reservoirs
- Geology, hydrogeology and structural features
- Topography
- Soil types, infiltration classification, general hydraulic properties and regolith soil storage
- Distribution of land degradation factors including soil salinity, waterlogging, soil acidity, soil structural decline, soil compaction and non-wetting soils (Williamson *et al.* 1997).

The list indicates there is consistent evidence that the major factors in determining the mobilisation of salt in landscapes are knowledge of changes in the hydraulic head of the various ground water systems as they respond to changing rates of recharge and discharge, and the amount of salt in the ground water. The impact of soil degradation on rooting depth, the plant types used in agriculture, and their effective rooting depth, are of major significance in producing changes to rates of drainage below the root zone, and hence recharge of the ground water aquifers.

Irrigation areas are generally located in landscapes with low relief. The excess irrigation water develops a ground water mound beneath the area from which the horizontal flow of ground water is restricted. Consequently, drainage into regions outside the irrigated area is quite limited. If in the surrounding dryland agriculture areas, there is some ground water recharge due to the removal of native vegetation, the drainage of ground water from the mound of water under the irrigation area can be further retarded.

The source of water-mobilising salt in irrigated lands is the excess water applied during irrigation, but in non-irrigated areas it is water normally used by the native deep rooted perennials. In the Western Murray Basin, recharge under native mallee vegetation is less than 0.1 mm/yr, however after clearing, the mean recharge rates are between 5 and 30 mm/yr (Allison *et al.* 1990). As the watertable in much of the Western Murray Basin is more than 30 m below the land surface, there is a considerable delay in the response of the aquifer to increased recharge. Analysis shows that while much of the area has been cleared for more

Box 1.2 Accumulation of salt in the landscape.

Where an arid basin is inundated by sea water and the sea retreats, evaporation incorporates the ocean salts in sediments and soils, and sodium sulfate may be enriched relative to sodium chloride. Both salts are neutral but in the presence of soil, sodium partly becomes an exchangeable cation that promotes alkalinity. Soils rich in neutral salts have been variously termed saline, white alkali or *solonchak,* while soils with sodium carbonate and exchangeable sodium are sodic, black alkali or *solonetz*. There are a range of definitions for sodic and saline soils depending on the country in which they are located. Some are more explicit than others with regard to the problem they are attempting to describe and define. Descriptions of salinity and related problems are generally presented from a soil description viewpoint rather than from an attempt to interpret the characteristics which are related to specific crops or species.

In drier areas, evaporation will leave the ions as alkali or carbonate salts. Sea water has an osmotic pressure of 24 atmospheres and contains nearly 35 g of ions per litre of which 55.5% is Cl, 7.7% SO_4, and 31.7% Na. The remaining 5.1% are mostly Ca and Mg. The neutral salts, NaCl (table salt) and Na_2SO_4, are highly soluble, 357 g/litre and 48 g/litre respectively at 0°C. During evaporation, sulfates precipitate much earlier than chlorides, and this facilitates salt separation (Jenny 1980). During weathering of rocks, calcium and magnesium which are slightly soluble in water tend to remain on the site whereas carbonates and bicarbonates of sodium and potassium are leached and transported.

than 50 years, the watertable is currently only rising in areas where it is less than 40 m below the ground surface. The development of salinisation is slow and so is the reversal of the process. The calculations indicate that unless land management is changed and new strategies incorporated, the Murray River will continue to steadily increase in salt for the next 200 years. In Western Australia, rates of rise of watertables in smaller catchments are much faster (Peck and Williamson 1987).

Ground water recharge beneath native vegetation in south-western New South Wales was less than 0.33 mm/yr in an area with 330–250 mm mean annual rainfall on aolean soils beneath mallee, open belah/rosewood woodland. Following clearing, recharge beneath annual agricultural crops and pastures increased between one and two orders of magnitude (20–40 mm/yr on sandy soils and 5 mm/yr on loamy soils). The increase in recharge leads to development of dryland and river salinisation since clearing of native vegetation, and its replacement with annual agricultural crops and pastures, results in an increase in the amount of water moving below the plant root zone. The increase in drainage percolates slowly through the unsaturated zone. When the extra drainage reaches the watertable, the watertable begins to rise. Because the river level remains unchanged, this causes an increase in the potential metric gradient towards the river and hence an increase in ground water discharge to the river. If the ground water is saline, this will lead to increases in river salinity. If the rising watertable enters the plant root zone, evaopotranspiration will concentrate salts causing land salinisation. Indications are that the period for response is 50 to 100 years (Allison and Hughes 1983; Cook and Walker 1989; Allison *et al.* 1990; Kennett-Smith *et al.* 1990; Walker *et al.* 1990; Cook *et al.* 1997).

Salinisation of water resources is a major problem. In Western Australia, most of the rivers and wetlands in agricultural areas have become saline (Schofield *et al.* 1989). Halting or preventing salinisation is dependent on establishing a hydrological regime similar to that operating before clearing. This can be achieved by establishing tree plantations in conjunction with other measures. Trees can be planted either on recharge areas (usually up slope) or on the saline discharge areas (Schofield *et al.* 1989). Recharge areas are more likely to be more suitable for commercial forestry than discharge areas where only a restricted number of tolerant trees are able to grow. The limitations of salt and waterlogging tolerance can be partially overcome by specific genetic selection and tree improvement.

In Australia, the major areas of non-irrigated saline soils are in the southern half of the continent, and have an associated impact on the surface water resource (Williamson 1990). Frequently, in the valleys the best agricultural land is where the saline soils develop. Scalds (exposed naturally saline subsoils) are located primarily in regions receiving less than 400 mm rainfall per year where grazing of natural vegetation is the dominant agricultural practice. The Murray–Darling Basin supports 75% of Australia's irrigated lands, with the consequence that New South Wales and Victoria contain 99% of the area of saline irrigated soils.

Without artificial drainage, the areas of shallow ground waters are expected to increase in irrigation areas and create conditions leading to increased soil salinity. Within the next 40 years, shallow watertables could underlie 70 to 80% of the major irrigation areas. The situation could leave 15 to 25% of the land with salinity high enough to render the land totally unproductive. The extent of loss in productivity has been considered for a wide number of projects and shown to be high to very high (Ghassemi *et al.* 1995) and the costs of rehabilitation to be extreme.

Classification of soil salinity and sodicity

Successful forest tree plantation establishment requires classification of sites into appropriate groupings to allow for efficient site-specific management. Classification of soils and sites in normal rain-fed plantation areas is recognised as an important part of plantation establishment (for example, Gasana and Loewenstein 1984; Turvey 1987; Turner *et al.* 1990). The characteristics and the systems used are determined by the scale and the basic type of country being established. Classification of plantations has two elements. One is the routine classification of plantations or potential plantation sites as a means of efficiently managing the area and the second includes experiments to allow suitable application of results to routine plantation areas. That is, what are the characteristics of the area and what is the technical base for its management?

The usual definition of a 'salt-affected soil' (Loveday and Bridge 1983) is a soil which contains sufficient soluble salt and/or exchangeable sodium to interfere with the growth of most crop plants. The interference may, in the case of soluble salts, be by way of osmotic effects on water uptake, nutritional imbalance or toxicity caused by specific ions. In the case of sodicity, it is particularly due to associated adverse physical properties (surface crusting, reduced permeability to water and air, and increased mechanical resistance to root penetration). A wide range of conditions apply throughout the world with the complication of salinity, sodicity and waterlogging presenting a complex matrix (Yadav and Singh 1970; Loveday and Bridge 1983).

Based on experience from broadscale plantation establishment, classification of potential forest plantation sites in relation to salinity needs to take into account the levels of salts and their distribution within the soil profile. A consideration for classification is the high level of

Box 1.3 A classification system for selecting sites appropriate for forest plantation establishment in saline or potentially saline environments.

Classification of soils and areas for the purpose of forest plantation establishment requires a system specifically directed to tree crops since they are deep-rooted, long-lived, and there are changes from juvenile to mature. This is critical where there are a range of potentially deleterious factors, such as salinity and waterlogging, in relation to plant growth. Such a system needs to include characteristics which are identifiable in the field or with minimal laboratory work, recognising the problems of measuring a highly variable parameter such as salt. Further, the selected characteristics must have known quantifiable effects on tree survival and growth. The approach here is hierarchal providing more refined information at each level.

STRATA 1: BROAD ENVIRONMENTAL ANALYSIS

[Objective: To broadly categorise the types of landscape for plantation establishment.]

LEVEL 1 Incident precipitation in mm per year

High	Moderate	Low	Very low
800+	600–800	400–600	<400

LEVEL 2 Landscape characteristics

[Objective: To define the type of salinity development and future problems.]

Dryland recharge
Dryland discharge
Dryland scald/erosion
Irrigated
Other
(Extent of each type needs to be addressed.)

STRATA 2: QUANTITIES OF SALT IN THE LANDSCAPE [(1)]

[Objective: To determine the quantity of salt and distribution in the landscape.]

LEVEL 3 Average level of salt in the surface soil (0–20 cm)

	Non-saline	Slight	Moderate	Severe	Extreme
Categories:	<2 dS/m	2–4 dS/m	4–8 dS/m	8–16 dS/m	>16 dS/m

LEVEL 4 Average level of salt in the deeper soil (50–100 cm)

Categories:	<2 dS/m	2–4 dS/m	4–8 dS/m	8–16 dS/m	>16 dS/m

STRATA 3: TYPES AND PROPORTIONS OF SALT IN THE LANDSCAPE

[Objective: To determine the types of salt and hence potential effects on soil structure.]

LEVEL 5 Proportion of sodium in base elements (ESP, Exchangeable Sodium Percentage):

ESP	<6%	non-sodic
ESP	6–14%	sodic
ESP	>14%	strongly sodic

LEVEL 6 Proportion of chloride in anions (relative to sulfates and carbonates)

<30%	low chloride
30–60%	moderate chloride
>60%	chloride dominant

LEVEL 7 ***Soil pH (interpret in relation to chemical distribution such as carbonates) and potential nutrient availability***

<5.0	very acid
5.0–6.5	acid
6.5–7.5	neutral
7.5–8.6	alkaline
>8.6	very alkaline

STRATA 4: HYDROLOGICAL CHARACTERISTICS

Objective: Soil limiting depth or access to water for growth.]

LEVEL 8 ***Watertable depth***

<0.5 m	extremely shallow
0.5–1 m	very shallow
1–2 m	shallow
2–4 m	moderate
> 4 m	deep

LEVEL 9 ***Irrigation water or ground water quality*** [2, 3]

[Objective: Irrigation may only be used at limited times but water has to be suitable.]

EC	<2.5 dS/m	can be used for most crops, good growth expected
EC	2.5–7.5 dS/m	plants with low to moderate salt tolerance
EC	7.5–15.0 dS/m	moderate tolerance to salt
EC	15.0–22.5 dS/m	poor growth expected in most plots
EC	>22.5 dS/m	not suitable

STRATA 5: FACTORS INFLUENCING ROOT DEVELOPMENT AND GROWTH

[Objective: To assess soil characteristics to undertake appropriate management.]

LEVEL 10 ***Profile texture: Uniform; gradational; duplex (duplex may have root limiting factor)***

Structure: Strong, moderate, weak, structureless (e.g. sand)
Impeding depth layer classes (defines availability of soil for growth):
5 m, 2 m, 1 m, 0.5 m
Soil organic matter content
Bulk density: High soil strength (more than 1.6 g/cc) will limit growth
Essential nutrients (determines requirements for fertilisers): N, P, K

COMMENTARY

The system outlined above attempts to address the factors considered important for commercial tree growth from a number of studies and places them into a logical sequence. Not all will be of relevance in each situation but it would be advisable to address each strata.

(1) Outlined in Marcar *et al.* (1995).
(2) Irrigation water quality has been described in ANZECC (1992) guidelines.
(3) Anon (1980).

variability of salt within the soil within the landscape (Conacher 1975). In addition to these factors, there is the broader issue of the quantity and seasonality of rainfall, considering that in most cases, salt will not accumulate in the soil where there is sufficient precipitation and water movement into the soil to allow leaching.

The classification system needs to take into account the required plantation management inputs when attempting to grow trees on saline or waterlogged sites. The saline conditions may have developed through a number of processes such as irrigation which may be ongoing. The site characteristics need to be identifiable and definable at a level to allow for the application of research and other technical information and for the utilisation of genetically improved material. In addition to the usual establishment techniques involving soil preparation, weed control, and nutrient manipulation, there is the further factor of irrigation. Where irrigation water is saline, there is the additional factor of modification of the existing saline environment, either adding to the salt load or assisting in the leaching of salts from the profile. To allow for appropriate development of planting material and efficient establishment techniques, a classification system for sites is outlined in Box 1.3. When characterising sites it needs to be kept in mind that soil conditions suitable for plant growth deteriorate as salts build up in the soil or the period of waterlogging increases.

The system of classification is broadly hierarchal allowing decisions to be made at each level. For example, if at Level 1 (Box 1.3), rainfall was 450 mm and at Level 2 it was dryland erosion, it would be of doubtful value for commercial products. If at Level 3, salt is found to be 15 dS/m, the area may be rejected from further analyses as too high to manage. If at Level 3, salt was moderate but at depth it was severe, it may be rejected from further analyses. Alternatively, the soils may be found to be acceptable, but at Strata 4, the bulk density may be found to be high with a shallow watertable leading to a decision to use mounding rather than ripping as an establishment technique. The system is established for decision making on site selection and when selected, how the site is to be managed.

Salt-affected soils are evaluated by measurements of their salinity and sodicity (Rhoades and Miyamoto 1990). Soil salinity refers to the presence of excessive levels of dissolved inorganic solutes, and is generally assessed as the electrical conductivity of the saturated extract (EC_e), since EC (electrical conductivity of a solution) is a practical index of total ionised solute concentration. Soil sodicity, which refers to the excessive presence of sodium, is generally expressed in terms of the exchangeable sodium percentage (ESP) or the sodium adsorption ratio of the saturation extract (SAR_e). The latter is more conveniently and accurately determined and these two parameters are closely related. The details of measurement are addressed in Chapter 2.

The proportion of the cation exchange capacity occupied by sodium ions (ESP) provides an index of the point at which soil structure and infiltration are affected (see Fig. 1.7). Sodic soils are categorised as those with an ESP from 6 to 14%, strongly sodic soils are those with an ESP of 15% or more. Soils with a high ESP are typically structurally unstable. A key aspect of SAR is the effect on the infiltration of water by reducing hydraulic conductivity. An indication of this can be shown in Fig. 1.7 where the effect of exchangeable sodium (as a proportion of total cations) on hydraulic conductivity is shown.

The classification system in Box 1.3 helps to identify areas suitable for planting and to identify stresses which may be imposed on trees. As a field example, such stresses have been identified in Western Australia and are as follows:

High soil salinity: total soluble salt in surface 20 cm of soil, typically in excess of 0.1% by weight and sometimes more than 1%.

- Waterlogging due to high watertables less than 0.5 m from surface (may be seasonal).
- Drought in summer and winter rainfall (Mediterranean climate).

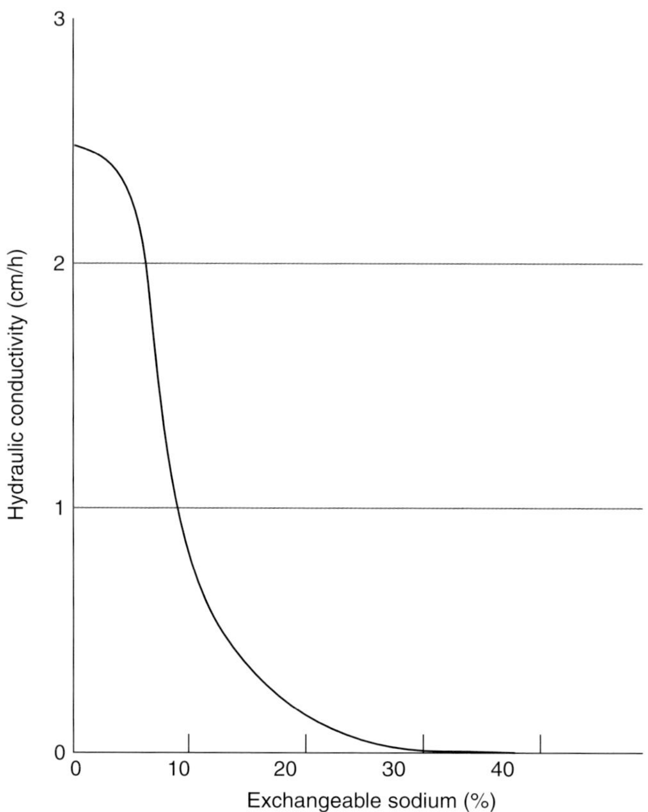

Figure 1.7 The effects of increasing exchangeable sodium as a percentage of exchangeable calcium, magnesium, potassium and sodium on soil hydraulic conductivity of a clay loam under acid and alkaline conditions (Martin *et al.* 1964).

- Low soil fertility.
- Hardpans preventing root growth below 0.5 m.

These factors are suitably reflected in Box 1.3.

Soil salinity is one of the most spatially and temporally variable properties of soils (see Table 1.8), and its variation within an area is generally much greater than the analytical errors. Thus the reliability of analytical data for salinity appraisal is often limited by sampling error (Rhoades and Miyamoto 1990). Soil description for many parameters is by inference and in the case of saline characteristics may be considered to be a gross and static indicator. In relation to plants, a better measure is a direct measure of salt in soil solution but in most cases this is impractical.

With regard to selection of tree species for the establishment of commercial crops, the factors in Box 1.3 relate to the issues of waterlogging, osmotic effects affecting water relations, toxicity or imbalances in relation to cations (sodium), toxicities or imbalances of anions (especially chloride and sulfate), and nutritional impacts due to deficiencies or reduced availability (due to high pH). These factors have been approached and measured differently

Table 1.8 Variation in soil properties with time (Jain *et al.* 1985).

		Month		
		October	**January**	**June**
Site 1	Medium salinity EC_e (dS/m)	4.5	11.5	26.0
Site 2	High salinity EC_e (dS/m)	20.0	25.5	36.0
Site 3	Very high salinity EC_e (dS/m)	27.0	33.5	40.0

in different regions, and there needs to be a more uniform characterisation of the effects and sites to optimise plantation site selection, establishment and growth.

Such a classification (Box 1.3) indicates key characters for commercial tree growth as outlined in the strata. Most reported studies have only provided information on parts of this proposed system and some other information has been inferred. Critically, it provides a framework for selecting plants by asking a series of questions, that is, can the plants tolerate high sodium, low oxygen levels, high chloride, reduced osmotic potential and/or a mixture of these. It needs to be recognised that any classification system developed to select areas for specific crops only provides a static index which may not be a good indicator of conditions actually controlling plant growth over time. Such is the case in salinity and sodicity where the plant roots are in contact with soil solution which is controlling plant growth but not readily classified in the field (Naidu *et al.* 1995).

The salinity of surface and ground water can also be conveniently measured with a portable conductivity meter. As the method of measurement will affect the final result obtained, it is critical that the system to be used is clearly stated.

Concluding comments

Salt accumulation in soils is an increasing problem throughout the world's productive agricultural lands. Trees provide one means to ameliorate and reverse this trend. However, the economic growth of trees requires a system of recognising the level and type of salt in the soil profiles in order to be able to select the most appropriate sites. Consideration of level of salinity in the field has to deal with the problem of extreme spatial and temporal variability.

Internationally, salt problems vary with geography, the main factors being differences in salt types and quantities. While there is considerable variation in salinity, the land management problems are fairly similar.

Consideration of site and species interactions for the development of plantations is important and a critical area for future research. Suitable classification of sites is critical for successful development and a system has been outlined.

CHAPTER 2

TREE CROP PHYSIOLOGY

Issues

Plants in saline environments are subject to a range of stresses including water deficits and elemental toxicities. Halophytes have a number of adaptive mechanisms to tolerate salt, but non-halophytes, being the main productive forest tree species, are only moderately tolerant and primarily use avoidance mechanisms. There is substantial variation between and within species in tolerance mechanisms to salt and these need to be understood, further developed, and utilised.

Salt tolerance in natural environments

Natural vegetation in saline habitats is often sparse, and highly saline areas are often barren with the exception of communities such as mangroves. Salt restricts growth of plants over larger areas of the Earth than any other inhibitory substance encountered in the natural environment (Epstein 1972). Some areas, such as on the coast, are affected by salt spray and form particular types of communities (Boyce 1954). Plants which grow naturally in saline habitats all possess special adaptive features and collectively are called halophytes (salt plants). However, this term does not refer to specific botanical taxa but rather to various physiological attributes. Two broad systems are used to define halophytes and are based on either the internal salt relations and salt tolerance mechanisms of the plant or on the salt content of the environment (Tal 1985). Analysis of the various forms of the adaptations of these plants may be of assistance in selecting physiological characteristics for commercial planting and production in saline areas.

Saline environments pose two distinct physiological threats to plants, namely desiccation and toxicity. They are derived from two unique features of the environment: low osmotic potential and high concentrations of sodium chloride and other ions which can be toxic to plants. There are additional factors which have impacts on plants in saline environments as the concentrations of essential elements may be low, and the presence of sodium may lead to deterioration of soil structure and reduced aeration.

Water relations are a critical aspect. There may be difficulty in plant water uptake, even when there is plenty of water in the soil, as a result of high salt causing low osmotic potential. That is, the plants must have very high internal salt concentrations to prevent osmotic desiccation because water would move osmotically from the plant cells into the substrate (soil) rather than in the other direction. Thus in terms of water relations, to cope with low external osmotic potentials, plants have to maintain potentials which are lower still (Box 2.1). This adjustment is critical in roots as this is the organ most exposed to the saline environment. Osmotic effects are likely to occur when plants are exposed to sudden changes in the root environment, for example, when seedlings are planted into saline soil or saline irrigation water is applied. Water stress symptoms may result within days or weeks (Marcar 1990).

Plants in saline environments need to be able to take up and accumulate salts in order to withdraw osmotically bound water from the soil. However, if these salts accumulate progressively throughout the life of the plant, the inevitable results are reductions in yield and eventually death (Larcher 1980). Even a high degree of salt tolerance fails when stress is prolonged and is continually increasing. Under such conditions, protective and compensatory mechanisms are vital — mechanisms that shield protoplasm from the effects of the stressor, or at least weaken and delay the effects (Box 2.2).

Halophytes are able to withstand high osmotic pressures in tissue fluids (15 MPa for *Atriplex* spp.) and the cellular enzyme systems endure high sodium concentrations (Flowers *et al.* 1977; Jenny 1980; Winter *et al.* 1981). Non-tolerant plants can react unfavourably to the negative water potentials generated by sulfates and chlorides. As soil water is removed by evapotranspiration, the resulting matrix and osmotic potentials are additive and water stress becomes inevitable. The focus of this discussion is on potentially commercial forest tree species. Given the full range of salinities, these species would be able to tolerate only moderate levels of salt salinities (2–10 EC_e in the soil) (Greenway and Munns 1980).

Box 2.1 Osmotic potential and water availability to growing plants.

Where pure water is separated from a salt solution by a membrane which is permeable to water but not salt, water diffuses into the salt solution. This process is osmosis and can be demonstrated in an osmometer (see figure below). In ecosystems, plant membranes and clay layers perform the function of the membrane. Eventually so much water will osmose that the 'weight' on the salt side will stop any further influx and the system is in equilibrium. Osmosis can be prevented by applying pressure on the salt side or by inducing a suction on the water side. The left arm is the sorption potential (equal to the negative osmotic potential, termed osmotic suction). This is the osmotic potential in megapascals (MPa).

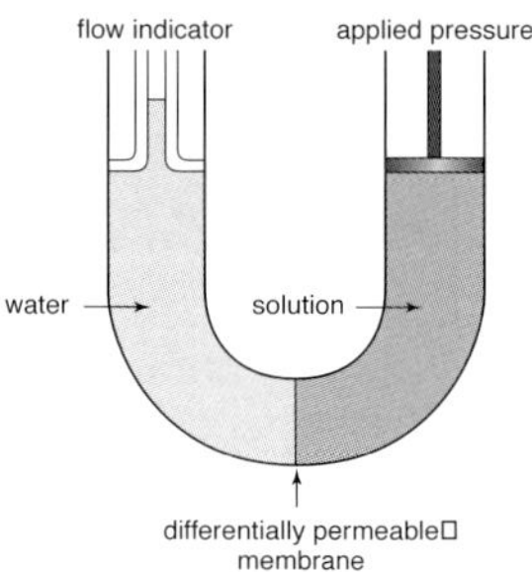

When the nutrient concentration in the cell sap of roots is higher than in the outside solution, water enters the cell, which is inflated and stiffened and is developing the turgor pressure (TP). Maximum turgor equals the osmotic pressure; as long as TP is less than osmotic potential, water keeps entering the cell. Since the osmotic potential is negative, the cell water potential is a negative quantity (Kramer 1969).

Water extracts of saline and alkali soils may generate potentials of more than 1 MPa which seriously interfere with good growth of crops. Roots penetrating a strongly saline soil and/or leaf tissue which is sprayed with salt may suffer plasmolysis, which is water flowing out of the cells causing plasma disruption and eventual death.

Above a ground water table, the major potentials of soil–plant ecosystems are gravitational (ψ_g) and sorptive (ψ_s), the latter being composed of matrix (ψ_m) and osmotic potential (ψ_o), or $\psi_s = \psi_m + \psi_o$. The two potentials are additive, provided the salt ions do not alter the soil matrix configuration. Tensiometers measure ψ_m but not ψ_o because their porous cups are permeable to solute. Likewise, pressure chambers for detecting water stress in plant materials respond to ψ_m only. The ψ_o is determined in expressed cell saps and displaced soil solutions by physico-chemical methods.

On a potential water scale, a level water surface at atmospheric pressure (0.1 MPa) has a potential of zero that denotes 'free water'. Air dry soils have large negative potentials, that is high suctions. Flow of water is from right to left.

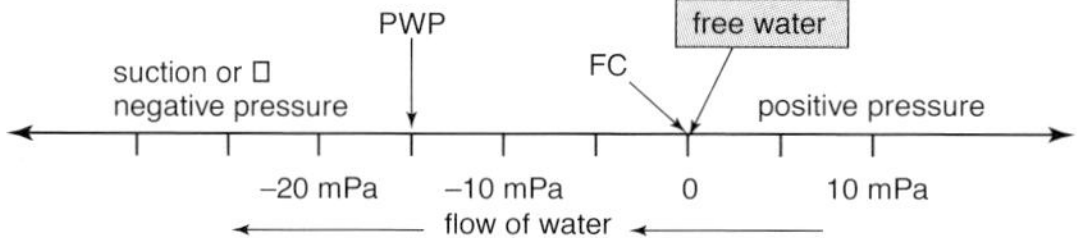

PWP = permanent wilting point
FC = field capacity

Box 2.2 Natural mechanisms for salt resistance in plants.

Salt resistance is the ability of a plant to withstand the presence of excess salts (especially related to the ions, sodium, chloride and sulfate) without serious impairment of vital functions. Plants in natural communities can resist or reduce the impacts of salt in the environment using a number of mechanisms. The mechanisms of reducing salt inputs are shown schematically (after Larcher 1980):

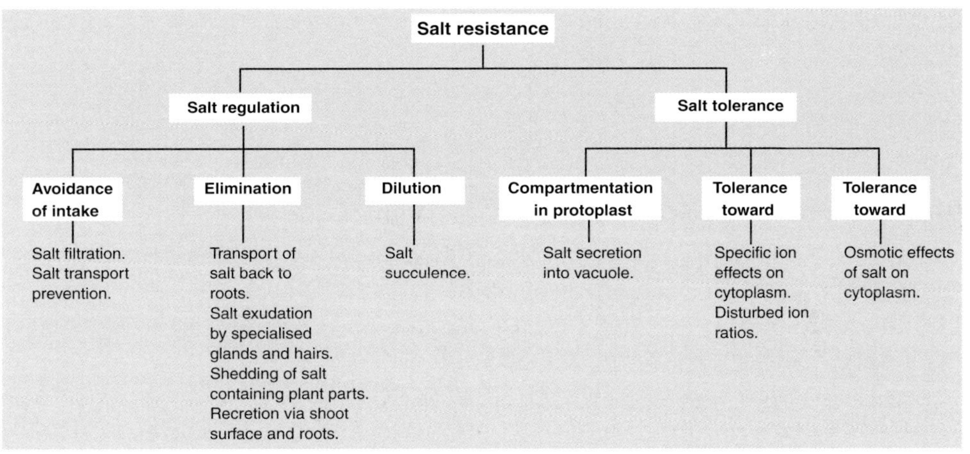

Salt regulation

The mechanisms of regulating the effects of salt include:

Salt filtration: Some mangrove trees (for example *Rhizophora*) greatly reduce the salinity of water in their conducting systems by ultra-filtration in the roots, thus stopping uptake of salts at the root surface.

Salt transport prevention: In some species, the transport of salt from the roots to the leaves is prevented. Salt ions (sodium in particular) are taken up by the roots, but they are retained there or in the trunk.

Salt elimination: Plants can rid themselves of excess salt by several mechanisms including downward transport through the roots and secretion through the roots, exudation and recretion at the surface of the shoot, and shedding of plant parts heavily loaded with salt. Salt-excreting glands and hairs actively eliminate salts, thus keeping their accumulation in the leaves within certain limits. Such glands are found within various mangrove plants (for example *Avicennia*), species of *Tamarix,* halophilic grasses such as *Spartina* and *Distichlis*. *Atriplex* accumulates chloride in vesicular hairs.

Succulence: Since the essential factor in the action of salt is not the absolute quantity but rather the concentration, a progressive accumulation of salts during the growing season can be compensated if the cells steadily draw in water, and in the process, they become considerably distended. The salt concentration in the cell sap then remains fairly constant. Chloride ions are responsible for the development of this type of succulence. Succulence is widespread among halophytes both in wet saline environments and dry regions.

Salt tolerance

Salt tolerance is a property of the protoplasm that enables it to be tolerated more or less well, depending on species, tissue type and level of vitality, and the changed ionic ratios associated with salt stress, and the toxic and osmotic effects associated with increased ion concentrations. Resistant protoplasts can survive 4 to 8% NaCl whereas salt-sensitive protoplasts are destroyed in solutions with as little as 1 to 1.5% NaCl.

Compartmentation in protoplasts: One factor in protoplasmic salt resistance is that salt ions taken into the cell are non-uniformly compartmented and most are stored in the vacuole, leaving the cytoplasm relatively low in salt. The cytoplasmic enzyme systems are less exposed to salt.

Tolerance: The cytoplasm of various species can vary with respect to specific ions and concentrations.

Plant growth and sodic soils

Although low osmotic potentials can be due to the presence of a soluble salt or other solutes, the term salinity commonly refers to solutions in which sodium ions predominate, the most prominent anion usually being chloride. However, in some salt-affected soils in drier areas, the osmotic effect is not so significant. Such alkali or sodic soils are those in which the cation exchange complex is occupied by sodium ions (6% sodium in soil context in Australia but often as high as 15% under some classifications), but which do not necessarily have much soluble salt in their soil solutions. The main constraints to plant growth in sodic soils include:

Poor soil physical conditions. Factors in sodic soils which impede root growth include poor aeration, high soil strength, stickiness when wet, and cracking soil surfaces. Sodic soils may also develop an impermeable layer, due to precipitation of calcium carbonate ($CaCO_3$) at depth which will severely restrict vertical root penetration.

Nutritional deficiencies and imbalances. Imbalances of elements cause more problems on non-saline sodic soils than on saline sodic soils because elemental concentrations in saline sodic soils are higher and cause direct competitive effects, which can be a problem for some tree species with high calcium demands. The pH of both saline and sodic soils can fluctuate significantly with season, sometimes by several pH units, and this will affect nutrient availability.

Toxicity of specific ions. A number of elements may occur in concentrations which are directly toxic to plants. Although sodium and chloride are the most prominent potentially toxic ions of saline substrates, other ions which are often found in saline soils may play important, often decisive, roles in the ecology and productivity of an area. Unregulated uptake of salts by plant roots and the inability of plants to place these salts within the vacuoles (which occupy most of the space of mature cells) of leaf cells can lead to leaf damage and reduced growth within weeks or months of exposure to saline conditions. For trees, chloride toxicity is considered more prevalent than sodium toxicity. High concentrations of magnesium may severely reduce root growth due mainly to induced calcium deficiency (Marcar 1989). Many subsoils in southern Australia may have high concentrations of boron leading to boron toxicity (Marcar *et al.* 1995). Most native tree species can tolerate quite high boron levels in their leaves and may not be individually affected.

A simple analysis of salt tolerance is not possible because of the interaction between tolerance and avoidance strategies, and between osmotic effects and specific ion toxicities (Greenway 1965). In general, salinity problems in forest trees in Australia are associated with low tolerances to specific ions, mainly sodium and chloride. Exclusion and compartmentation are the most common avoidance mechanisms in tree species. Exclusion involves discrimination against uptake of the toxic ion at the first membrane barrier of the root, while compartmentation occurs when potentially toxic ions are accumulated or compartmentalised in less vulnerable plant parts.

In assessing commercial plantation species for salt tolerance there needs to be an understanding of the mechanism of how salt is tolerated by the plant. Maintaining plant water status is a primary problem, and in addition there are problems of toxicity, mainly from sodium and chloride. Exclusion and compartmentation are two possible tolerance options, and the importance of these for individual species needs to be further analysed.

Measurement of soil salinity

Soil salinity may be measured in a number of ways, however the parameter chosen to relate salinity to plant tolerance must correlate closely with plant growth and yield (Maas and Hoffman 1977). Without specific ion effects, growth reduction is primarily related to the osmotic potential of the soil solution in the root zone. The degree of salinity of a soil is expressed in terms of the electrical conductivity (EC_e) of an aqueous suspension or extract usually taken from the active root zone in the soil. This measurement reflects the total concentration of ions in solution but it does not give any information about its composition in terms of specific ions.

Electrical conductivity is directly related to the concentration of soluble salts in the soil solution and within limits of osmotic potential, ψ_o, by the relationship, $\psi_o = -0.36\ EC_e$. Use of electrical conductivity is recommended because the saturation percentage is easily and reproducibly determined in the laboratory and is related to the field-moisture range of soils varying widely in texture. For many soils, the soluble salt concentration of the soil solution at field capacity (the soil water content after drainage of gravitational water has become very slow and water content is relatively stable) is about twice that at saturation. Salinity measurements obviously would be more meaningful if determined from soil solutions in the field-moisture range rather than from saturated solutions, however this is often impractical.

Electrical conductivity may be measured in the laboratory using either a 1:5 soil:water suspension ($EC_{1:5}$) or by measurement on a saturated extract (EC_e); see Box 2.3 for conversion of $EC_{1:5}$ to EC_e. The EC_e is considered to be more closely related to field conditions, however estimation is a much more tedious procedure (Shaw 1999). Most of the reports of data related to salinity are from $EC_{1:5}$ suspension.

The distribution of salt within the soil profile is important in plant response. Several studies support the hypothesis that plants respond to the mean salinity in the root zone while other studies indicate that the effective salinity level must be weighted in favour of the least saline zone. Such studies have found that plants can tolerate excessive salinity levels if a sufficient part of the root zone is relatively salt free. These aspects emphasise the problem of assessing salinity levels when dealing with deep-rooted long lived woody crops as the root distribution varies both spatially (depth) and over time.

Box 2.3 Multiplication factors for converting $EC_{1:5}$ to EC_e for different soil textures (Slavich and Petterson 1993).

The quantity of salt in the soil, soil salinity, is best described in terms of electrical conductivity (EC). The measure of electrical conductivity gives a general estimate of the quantity of salt present but not the type of salt. The best measure is the electrical conductivity of a saturated soil paste (EC_e) in units of deciSiemens per metre (dS/m). Electrical conductivity of a 1:5 soil:water extract ($EC_{1:5}$) is the alternative (and easier to determine) measure. $EC_{1:5}$ can be converted to EC_e by applying multiplication factors.

Group	Soil texture field test	Multiplication factor
Sands	Very little or no coherence and cannot be rolled into a stable ball; individual sand grains adhere to fingers.	17
Loams	Can be rolled into a thick thread, but will break up before it is 3–4 mm thick; the soil ball is easy to manipulate and has a smooth spongy feel with no obvious sandiness.	10
Clay loams	Can be easily rolled into a thread 3–4 mm thick but will have a number of fractures along its length; the soil is becoming plastic, capable of being moulded into a stable shape.	9
Light clays	Can be rolled to a thread 3–4 mm thick without fracture; plastic behaviour evident, smooth feel with some resistance to rolling out.	8
Medium clays	Handles like plasticine, forms rods without fracture, has some resistance to ribboning shear, ribbons to 7.5 cm or more.	7

While taking soil samples and measuring electrical conductivity provide a reasonable estimate of salinity, it is inefficient for extensive field assessment of areas. In the field, bulk soil electrical conductivity sensors or electromagnetic induction sensors may be used. Electromagnetic induction instruments (for example EM-31, EM-34, EM-38) for which there are variations in their penetration depths into the soil, are being developed for broadscale assessment and to be used in a range of areas (for example Norman 1989, 1990; Norman and Heath 1992; Beasley 1994). The EM-38 instrument has been used to survey large areas to ascertain the extent and level of salinity. The instruments measure apparent electrical conductivity (EC) of the soil profile to several metres but need to be calibrated against standard estimates of salt salinity. The EM-38 instrument operates through the induction of an electromagnetic field in the soil emanating from a transmitter coil located in one end of the instrument. This produces electrical currents in horizontal planes at various depths (Fig. 2.1). These current loops generate a secondary magnetic field that is proportional to the value of the current flowing within the loop (Corwin and Rhoades 1982). A fraction of the induced electromagnetic field from the current loops is sensed by the receiver coil located 1 m away at the other end of the instrument. This signal is then amplified and the ratio of the secondary to the primary magnetic fields is transformed into an output voltage that is linearly related to the apparent soil electrical conductivity (EC_a). This is measured as conductance in milliSiemens per metre (mS/m). The reading is an integrated measurement of the soil conductivity to a depth of about 3 m (depending on the

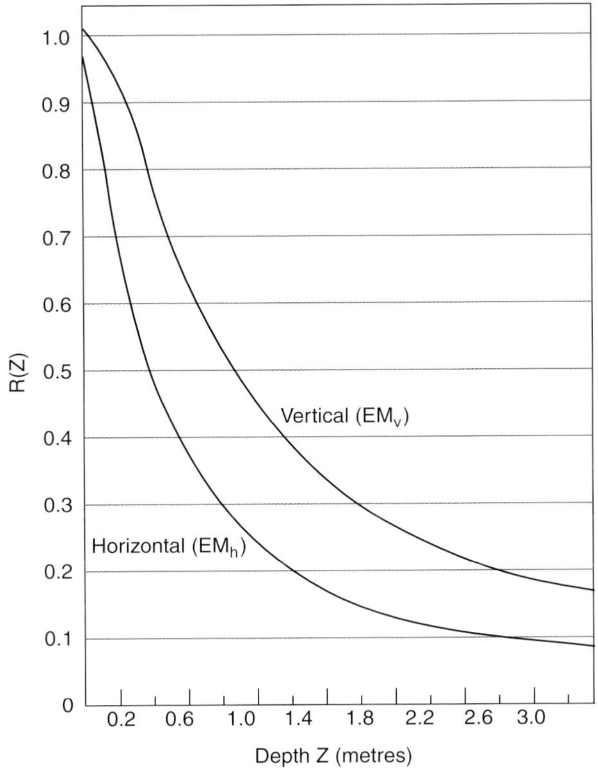

Figure 2.1 The cumulative relative contribution of soil electrical conductivity, *R*(Z), to the EM-38 reading below various depths, with the instrument held in a vertical (EM_v) and horizontal (EM_h) position (Norman and Heath 1992).

alignment of the coils at the time of the reading), although the instrument is most sensitive to conductivity in the top 0.6 m of the soil profile.

There has been a considerable amount of discussion about instrument calibration due to the effects of soil moisture, texture and salinity (Slavich 1990; Slavich and Petterson 1990). A standard calibration measure is the electrical conductivity of a saturated soil paste extract (EC_e) or a suspension (1:5). The calibration needs to be checked against various soil types due to the variations (for example, Fig. 2.2) (McNeil 1980; Cameron *et al.* 1981; Rhoades and Corwin 1981; Corwin and Rhoades 1982; Rhoades *et al.* 1989; Lesch *et al.* 1992). However, the method has proved valuable in providing bulk estimates of the highly variable property of salinity in relation to tree growth (Norman and Heath 1992; Bennett and George 1995). While it may prove too broad an estimate in assisting selection for tree improvement on the basis of salinity around individual trees, the method is of high value in terms of soil and site classification for broadscale plantations.

Mechanisms of uptake of elements

Plants take up inorganic chemicals from the soil solution and these include those essential for growth plus others, such as salt, which are non-essential or even toxic. There are a number

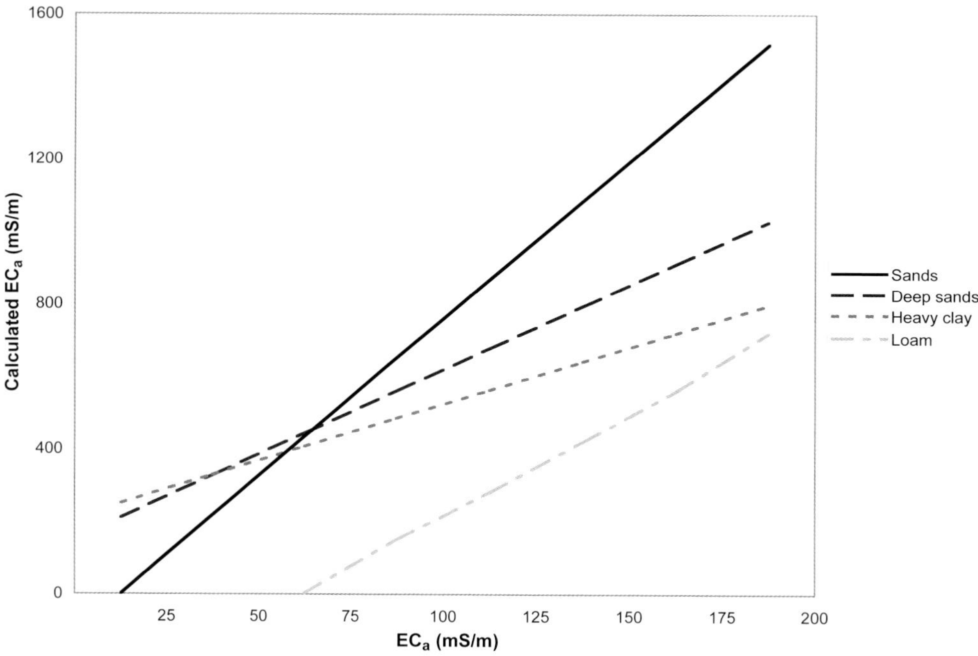

Figure 2.2 Comparison of different EM-38 calibration equations from south-west Australia showing the effect of soil type (from Bennett and George 1995).

of processes by which elements are taken into the roots and these are described as passive and active uptake mechanisms (Epstein 1972). Consideration of active processes has focussed on essential elements where plants are acquiring and accumulating nutrients from dilute solutions or exchange complexes. At relatively low concentrations of salt ions in solution, many plants can restrict the uptake of cations and/or anions. This is the single phase uptake. At higher concentrations, there is more rapid, less restricted uptake possibly related to mass flow, that is the uptake is equivalent to the quantity of nutrients in the water taken up in the transpiration stream and is less controlled. This is the second phase, hence the term 'dual mechanism' of uptake. Where trees are considered in terms of salinity, there is evidence that in the first phase they generally use exclusion while in the second phase of rapid uptake they utilise compartmentation of elements such as accumulation in bark (Sands and Clarke 1977). The concentration at which mechanism 2 occurs varies for each element.

Epstein (1972) discussed the dual mechanism of uptake using potassium as an example. The mechanism which affects the absorption of potassium over the usual (low) range of solution concentrations has several well-defined properties: it operates at very low concentrations of potassium, follows simple Michaelis–Menten kinetics, is specific for potassium and indifferent to the anion. If this mechanism were the only one to transport potassium into the cell, then even at much higher concentrations of potassium, the rate of its absorption would not exceed the maximal rate asymptotically approached at the highest concentrations used. In terms of carrier hypotheses, virtually all carrier sites available for transport of potassium are occupied at these concentrations, and raising the solution concentration still further cannot cause the rate to exceed the maximum corresponding to complete occupancy of all sites.

However there is more than one mechanism for absorption. Mechanism 2 which becomes operative at high concentrations, leads to a more rapid rate of uptake say at higher than 1 mM potassium. Mechanism 1 which operates at low concentrations differs entirely from mechanism 2. Mechanism 1 is ion specific whereas mechanism 2 has competition with other cations. Further, the form of the anion affects the mechanism of uptake. For example, sulfate suppresses potassium uptake in mechanism 2. Processes of type mechanism 2 have been demonstrated for chlorine, sodium, ammonium, boron, and other ions.

Few studies have been carried out on forest tree species in relation to mechanisms of uptake. In one study on Norfolk Island Pine (*Araucaria heterophylla* (Salisb.) Franco) (Truman and Lambert 1978), seedlings were grown in nutrient solutions to study increasing concentrations of sodium, potassium and chloride. The uptake indicated dual mechanisms of uptake for all three elements with exclusion at low concentrations. The concentration at which mechanism 2 became significant varied with the element. For example, potassium changed from mechanism 1 to mechanism 2 at about 5 mM solution, while for sodium this was about 100 mM and for chloride about 200 mM. That is, the species was excluding sodium and chloride up to relatively high concentrations. The evidence from several studies (Sands 1981; J. Turner and M. J. Lambert *pers. comm.*) indicates that the dual mechanisms operate for some eucalypt species (for example *E. camaldulensis*) where the change occurs at about 100 mM NaCl for both sodium and chloride mechanisms. At about 200 mM NaCl, there is rapid uncontrolled uptake of salt.

At lower salinities, uptake of salt occurs because there is a mass flow of saline solution towards the root as a result of the transpiration stream, and the root is imperfect at discriminating between saline and nutrient ions. Entry of salt may occur by leakage across the root membranes by poor selectivity of carriers, or can channel at some point along the radial pathway to the xylem or by leakage along an apoplastic pathway from the outside the root to the xylem. Selectivity between nutrient and non-nutrient ions is usually considerable, even in salt-sensitive species, but it may be insufficient to cope with a saline environment. The more or less efficiently that the plant uses water, the lesser or greater the flow of saline solution towards the plant for each unit of growth.

Exclusion of salt may be a sufficient protective system for low salinities and for the most sensitive crops where salt damage is almost entirely as a consequence of internal accumulation. If photosynthesis is reduced, productivity will be reduced. However exclusion may prejudice tolerance of higher salinities. Since the overall objective of salt exclusion is to limit the concentration in the leaves, then a rapid growth rate (dilution by growth) will achieve a result similar to restriction of uptake.

The consideration of such uptake mechanisms means that in the analysis of salt tolerance, the point of change from mechanism 1 to mechanism 2 is critical, particularly for sodium and chloride. In fact, the stable level of mechanism 1 and subsequently the rate of uptake of mechanism 2 will each have an effect on the characterisation of salt tolerance. When considering sodium and chloride for a species, it is highly probable that the change from mechanism 1 to mechanism 2 will be different for each element and if that change point for sodium is lower in concentration than the change point for chloride, the species may be more susceptible to damage from sodium. Toxicity probably occurs at a concentration higher than the change point. Where genetic improvement is involved, a species may be considered salt tolerant but there needs to be more detailed characterisation of such tolerance and the

Table 2.1 Schematic salt tolerance profile for *E.camaldulensis*.

	Sodium	Chloride
Change point from mechanism 1 to 2	Low	Very high
Plateau concentration of mechanism 1	Low	Moderate
Uptake rate in mechanism 2	High	Very high
Compartmentation	Moderate (accumulates in bark)	High (accumulates in bark)

processes involved. In characterising a species tolerance of sodium and chloride or otherwise, there are several points of change since, as the concentration of an element changes in solution, this includes the change point from mechanism 1 to mechanism 2, the plateau level of mechanism 1 (exclusion potential) and the rate of uptake of mechanism 2 (the rapid rate of uptake). In addition, there is the potential especially in older plants, for compartmentation. Schematically and in subjective terms for *E. camaldulensis*, a salt tolerant profile can be developed (Table 2.1).

Such a profile (Table 2.1) indicates a moderate tolerance to sodium, a high tolerance to chloride in relatively low concentrations and a greater sensitivity to chloride than sodium. As the tree matures it may be able to tolerate chloride better than sodium through compartmentation. In a saline situation, a tree would be most affected by the element to which it is least tolerant. Such a profile matrix (which could be extended for other characteristics) potentially represents the critical characteristics required for screening in species selection programs.

Salt uptake in field grown trees

Maintenance of water uptake and transpiration is critical for maintaining stand survival and growth. In commercial plantations, water may be a major limiting factor in long term viability. Water requirements at a stand level as indicated, for example, by transpiration rate, vary with plantation age and growth rate. Generally there is a peak of demand near crown closure followed by a decline in requirements with increasing age, a typical pattern being shown in Fig. 6.3. The growth requirement is one component of stand water balance and is outlined in detail in Box 6.2. Water use efficiency is an index of how much water is required to produce a unit of organic matter. It has been shown to vary with genotype, both between and within species and probably varies with age. There was a variation (by a factor of 2) between clones of *E. grandis* for a given amount of water transpired in that one clone produced about twice the amount of organic matter in small plants (Olbrich *et al.* 1993). Whether these estimates relate to field conditions has not been tested, however it appears selection is possible for water use efficiency. In calculations below (Table 2.2), a standard figure for water use efficiency has been used to estimate water used based on known organic matter production.

Growth in field grown trees does not appear to be affected by salt within a low range of salinities and this is probably related to uptake mechanism 1. When salinity increases there is rapid decline in productivity. This decline is probably related to a rapid accumulation of salt arising from high uptake in some cases through less regulated uptake, possibly as mass flow. While mass flow is a consideration, it is difficult to determine under field conditions

Table 2.2 Estimates of actual uptake (kg/ha/yr) and potential uptake (kg/ha/yr) from mass flow for sodium, potassium and chloride for field grown trees.

Species	Actual uptake of Na	Potential uptake of Na	Actual/ Potential (%)	Actual uptake of K	Potential uptake of K	Actual/ Potential (%)	Actual uptake of Cl	Potential uptake of Cl	Actual/ Potential (%)
Casuarina cunninghamiana[1]	2.9	325	0.9	33.7	93	36			
Casuarina glauca[3]	35.8	5320	0.7	194.5	598	33			
E. camaldulensis[1]	6.8	346	2.0	28.0	99	28			
E. camaldulensis[2]	27.0	9000	0.3	56.6	1310	43	299	7196	4.2
E. camaldulensis[3]	38.5	6284	0.6	71.6	707	10			
E. camaldulensis[4]	5.1	1517	0.3	91.0	239	38	73	2410	3.0
E. grandis[1]	10.9	560	1.9	22.1	160	14			
E. grandis[3]	43.2	5652	0.8	64.4	636	10			
E. globulus[3]	104.4	6478	1.6	111.9	729	15			
E. globulus[4]	48.0	1320	3.7	69.2	211	33	68	1920	3.5
E. occidentalis[3]	76.8	5211	1.5	86.8	586	15			
E. saligna[1]	12.2	547	2.3	23.9	155	15			
P. radiata(1)	2.1	308	0.7	28.7	88	33			
Poplar 70/51[1]	3.9	256	1.6	28.8	73.2	59			
Poplar 65/21[1]	3.2	360	0.9	33.4	103	32			

[1] Stewart *et al.* (1988); [2] Kube *et al.* (1987); [3] Boardman *et al.* (1996a); Hanna *et al.* (1992); [4] J. Turner and M.J. Lambert (*unpubl. data*).

but has been undertaken in a number of studies (Ballard and Cole 1974; Prenzel 1979; Turner 1982). In such studies, estimates of actual uptake and average soil nutrient solution concentration are made. A comparable analysis has been made where actual uptake in young trees has been assumed as the total elemental accumulation, while potential uptake is estimated using calculated transpiration and the average concentration in the applied effluent or soil solution. That is, the potential uptake is assumed to be the product of the quantity of water transpired by the stands and the elemental concentration of soil solution.

All species studied were actively limiting the uptake of sodium with the actual uptake being less than 3.7% of potential uptake. The quantities of sodium potentially available ranged from 256 to 9000 kg/ha/yr. The estimates of chloride indicate that this element is excluded by *E. camaldulensis* and *E. globulus* as only 3 to 4.2% of potential uptake was actually taken up. Potassium is also limited but to a lesser degree and appears to be more species related. Actual potassium uptake ranged from 22.1 kg/ha/yr in *E. grandis* to 194.9 kg/ha/yr in *Casuarina glauca*. The actual uptake as a percentage of potential uptake ranged from 10 to 59%. Under these conditions, unregulated uptake does not appear to be a major factor. The evidence is that most studies on salinity processes in forest trees are carried out where soil solution concentrations are low and at which only uptake mechanism 1 is operating, that is salt levels in the plant are being limited by active exclusion. Mechanism 2 will lead to rapid uptake of specific ions leading to damage and reduction in growth. The concentration at which mechanism 2 is operative varies with species and the specific ion, but in most studies is not defined and not achieved in most field studies.

Breeding plants for tolerance to environmental stresses such as salinity presents situations in which visible characteristics do not provide sufficient information for plant breeders and for which yield is not an efficient index of potential parent lines (Yeo 1994). Part of this is due to the complexity of the pressures placed on the plant rather than due a single identifiable agent. Tolerance usually has visible expression only in stressful conditions and so it is necessary to locate suitable field sites for screening in trials to assess both potential parent and breeding populations. In the early stages of selection where both plant material and soils are highly variable, the characterisation of individuals can be done with only a low level of confidence. Characteristics which provide objective quantitative information over short time periods can improve the processes and information context, and reduce interference due to environmental effects. Some physiological criteria may be used (photosynthesis, polyamine) and these may be specific enough to isolate individual traits.

With avoidance mechanisms, the process of keeping low levels of salinity in leaf laminae, the mechanisms are visually indistinguishable: the only visible features being damage. It cannot be deduced from appearance whether plants growing without damage in saline soils possess genetic information for one or more tolerance mechanisms or are simply the most vigorous.

Salt uptake through foliage

The level of resistance of a species to salt damage will depend on whether the salt is absorbed by roots from saline soil or whether the salt is absorbed through the leaves. Trees have been planted in locations affected by high foliar salt such as near the sea or near highways where de-icing salts are used. A tree may be well adapted to avoid accumulation of toxic ions in foliage from soil by exclusion and compartmentation, but be poorly adapted to avoid intake of toxic ions through the foliage. The protection mechanisms of the foliage tend to be

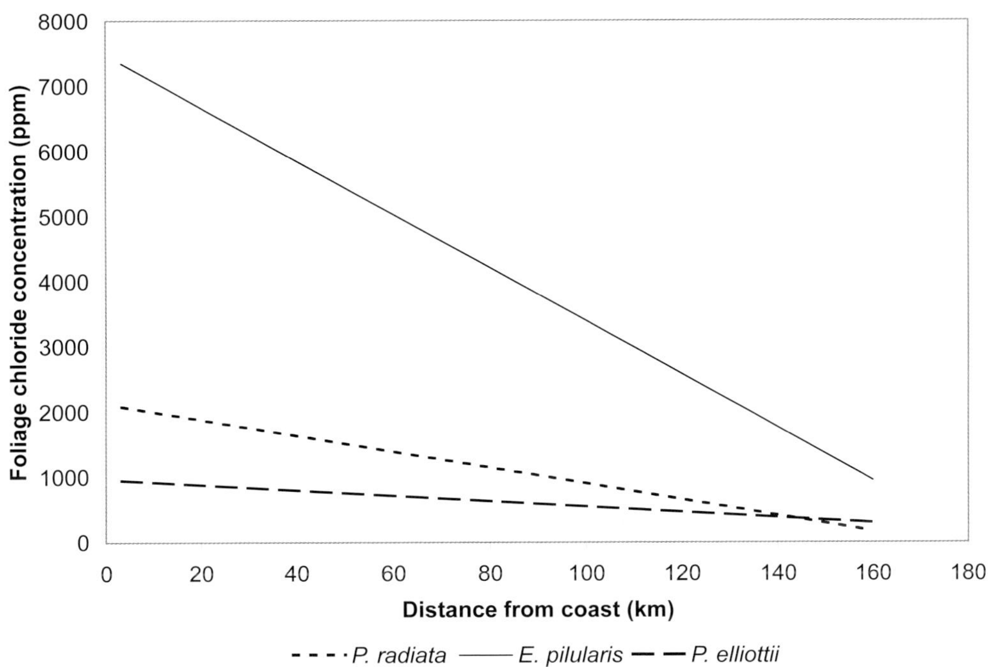

Figure 2.3 The relationship between foliage chloride concentrations and distance from the sea in *E. pilularis* and *P. elliottii* trees in forests in Australia.

physical barriers such as wax coatings. Effects of toxic levels of salt in foliage need to be identified and separated from uptake and damage through roots.

The main focus of this book is the development of commercial plantings in environments where the soils are high in salt. However stresses can be placed on trees due to uptake of sodium and chloride through foliage. Assessments have been carried out for ornamental species in areas with increased salinity in environments (Karschon 1958; Holmes and Baker 1966; Davison 1971; Lumis *et al.* 1973; Dirr 1974, 1976; Shaybany and Kashirad 1978; Dowden and Lambert 1979; Townsend 1980) and sources of salt can be spray from saline solutions such as aerial irrigation and seawater spray. Species differ in their ability to tolerate salt in foliage, for example eucalypts are relatively intolerant of windborne coastal salt, while *P. radiata* and *P. elliottii* are more tolerant and some species, such as *Araucaria,* are very tolerant. Similar effects can occur due to the splashing of de-icing salts applied to roads (Holmes 1961; Holmes 1966). In natural situations, concentrations of sodium and chloride in foliage of forest trees are found to increase greatly when growing in proximity to the ocean, and there is a strong inverse relationship between distance from the ocean and sodium and chloride concentrations in foliage which is high in immediate proximity to the ocean with a very rapid decline inland (Kurauchi 1956; Edlin 1957; Karschon 1958; Turner 1972; Turner and Kelly 1973). Species vary and this is indicated by comparison of chloride concentrations of three species planted in NSW, Australia (Fig. 2.3). A generally similar pattern occurred for sodium.

Karschon (1958) showed a similar relationship for *E. camaldulensis* in Israel (Fig. 2.4) indicating very great decline in foliage chloride up to 3 km from the ocean and then a much slower decline. Sodium was much more variable and appears to be more affected by salt variation.

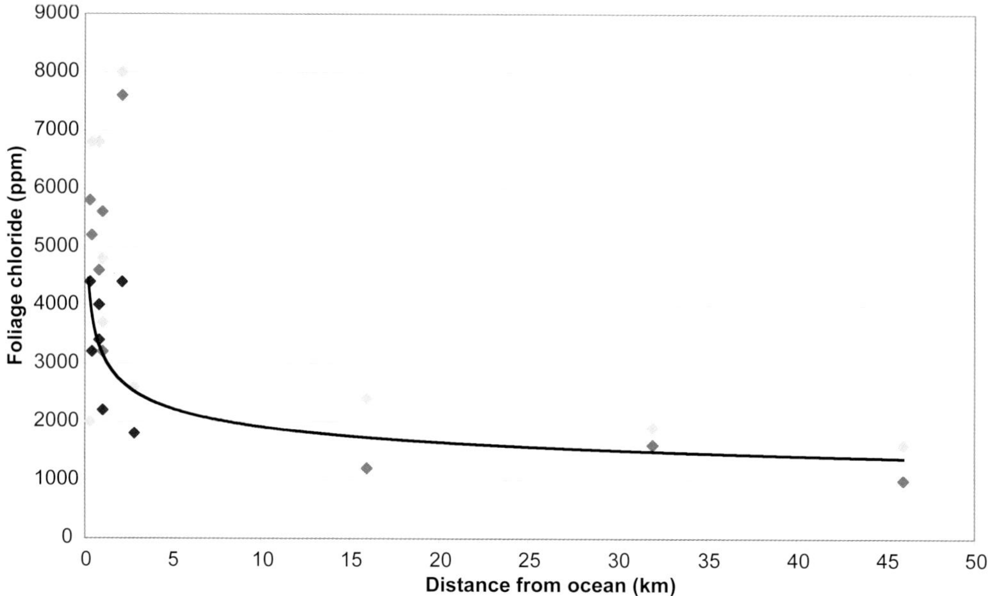

Figure 2.4 Variation in foliage chloride (ppm) in *E. camaldulensis* with increasing distance from the ocean (after Karschon 1958). The variation in intensity of the diamonds reflected different levels of drainage.

Uptake of salt through foliage can be increased by the presence of surfactants (wetting agents in solution) which remove protective foliage waxes from otherwise salt tolerant species. Such an effect was demonstrated in a study of coastal plantings of Norfolk Island Pine (*Aruaucaria heterophylla* (Salisb.) Franco) where the surfactants were discharged in sewage from outfalls and were present in aerosols in spray which reached trees due to wind and wave actions (Dowden and Lambert 1979). Norfolk Island Pines are well adapted to growing under seafront conditions as the roots are able to effectively exclude salt and can grow in up to 80% sea water (Truman and Lambert 1978). Salt in sea spray enters the needles through the stomata and the uptake of salt to toxic limits is avoided since the relatively tough and impermeable cuticle is particularly resistant to salt (chlorides of sodium, potassium, calcium, magnesium) (Grieve and Pitman 1978). However, toxic limits were exceeded in trees planted near coastal sewage outflows (Dowden and Lambert 1979) since the presence of surfactants lowered the surface tension of sea spray sufficiently to promote stomatal entry of salt (Dowden *et al.* 1978; Grieve and Pitman 1978). Similar effects have been noted in other areas and species where surfactants are present from ocean sewage outfalls (Anderson *et al.* 1981; Gellini *et al.* 1985; Astorga *et al.* 1993; Bussotti *et al.* 1995). This form of injury from foliar uptake of salt is not usually a major problem in commercial plantations.

For commercial production, foliage uptake and damage will be of only minor concern and would mainly occur if spray irrigation systems were used.

Tolerance of plants to salt

Crop salt tolerance is usually expressed as the decrease in yield expected for a given level of soluble salts in the root medium compared with the yield under non-saline conditions (Maas

Table 2.3 General crop sensitivity to salinity (Maas and Hoffman 1977).

Crop	Scientific name	Threshold salinity of saturation extract (dS/m)	Decrease (%) in yield at soil salinities above the threshold (dS/m)
Sensitive crops			
Bean	*Phaseolus vulgaris* L.	1.0	19
Strawberry	*Fragaria* spp.	1.0	33
Almond	*Prunus dulcis* (Mill.)	1.5	19
Plum, prune	*Prunus domestica* L.	1.5	18
Orange	*Citrus sinensis* (L.)	1.7	16
Moderately sensitive crops			
Radish	*Raphanus sativus* L.	1.2	13
Clover, strawberry	*Trifolium fragiferum* L.	1.5	12
Clover, ladino	*Trifoliun repens* L.	1.5	12
Grape	*Vitis* spp.	1.5	9.6
Corn	*Zea mays* L.	1.7	12
Tomato	*Lycopersicon* spp.	2.5	9.9
Rice, paddy	*Oryza sativa* L.	3.0	12
Moderately tolerant crops			
Sudangrass	*Sorghum sudanense*	2.8	4.3
Ryegrass, perennial	*Lolium perenne* L.	5.6	7.6
Wheat	*Triticum aestivum* L.	6.0	7.1
Tolerant crops			
Bermuda grass	*Cynodon dactylon* (L.)	6.9	6.4
Cotton	*Gossypium hirsutum* L.	7.7	5.2
Barley	*Hordeum vulgare* L.	8.0	5

and Hoffman 1977). However such an expression of salt tolerance is not an absolute value, rather it is a relative value based on cultural conditions under which the crop is grown. Absolute tolerances that reflect predictable inherent physiological responses by plants cannot be determined because many interactions among plant, soil, water and environmental factors influence the plant's ability to tolerate salt. A large number of procedures and methods have been used to determine salt tolerance and measure plant responses, and many of these are not readily comparable. The above definition of salt tolerance was developed specifically for crop plants but is also applicable to plantation trees where levels of organic matter accumulation are expected over time.

Although salinity physiologically affects plants in many ways, overt injury symptoms seldom occur except under conditions of extreme salinisation (Maas and Hoffman 1977). Salt-affected plants usually appear normal, although they are stunted and may have darker green leaves, which in some cases are thicker and more succulent. Woody species are an exception since toxic accumulations of chloride and sodium cause leaf burn, necrosis and defoliation. Although most plants respond to salinity as a function of the total osmotic potential of soil water without regard to the salt species present, some herbaceous plants and most woody species are also susceptible to specific ion toxicities. In some cases, salinity induces nutritional imbalances or deficiencies resulting in decreased growth and plant injury which cannot be related to osmotic effects alone, for example where low calcium or magnesium levels occur when sulfate levels are excessive. The relationship between osmotic potential and the soil solution and crop yield is invalid under conditions in which specific ion effects are significant. Relative salt tolerance in crops is shown as a demonstration in Table 2.3 (Maas and Hoffman 1977).

Processes limiting growth of non-halophytes in saline soils have been analysed (Munns and Termaat 1986). Leaf growth was considered to be more sensitive to salinity than root growth, so initial focus was on processes that limit leaf expansion. Effects of short-term exposure (days) were considered separately to long-term exposure (weeks to years). A factor in the short term is probably the water status of the root and it is suggested that a message from the root regulates leaf expansion. A limitation to growth in the long term may be the maximum salt concentration tolerated by fully expanded leaves. If the rate of death approaches the rate of new leaf expansion, the photosynthesis area will eventually become too low to support continued growth.

Nutritional or elemental problems will occur as a result of or in conjunction with elevated soil salinity. Symptoms related to high or low levels of elements may assist in interpreting specific problems. Toxic levels of chloride in *E. camaldulensis* led to small necrotic patches on older leaves, and these enlarged rapidly prior to their premature shedding. Newly expanded leaves developed chlorotic symptoms and became markedly thickened (Thomson 1988). However, forms of necrosis and chlorosis will arise from other elements and some symptoms are shown in Box 2.4.

Salt accumulations in plantations

Where salt is taken up in increasing quantities, the question arises as to how forest tree species manage it. Salt tolerance in forest trees can either be due to restriction of uptake of ions or compartmentalisation of them. Such processes appear to vary according to species, age and specific ions, although the data are very limited. Trees need to be at the sapling stage to assess their specific tolerance to salt. Notwithstanding the numerous mechanisms of salt tolerance, the enhanced ability to exclude sodium and especially chloride ions from roots or shoots is the most important mechanism operating in salt tolerant lines of woody species (Allen *et al.* 1994), with the next most important being compartmentalisation. Evidence from *Eucalyptus* spp., *P. radiata* and *Casuarina* spp. indicates that there are various mechanisms of salt tolerance and resistance in plantations and that they change over time. The actual concentration of salt in the root environment is critical. It appears that most studies are at relatively low concentrations, for example mechanism 1 of Epstein (1972) where species can control (restrict) intake of cations and/or anions to the plants. The critical point from mechanism 1 to mechanism 2 varies with species and specific ions, but it occurs when changeover concentrations move above the point where less controlled mass flow uptake occurs. Trees vary in the way this is handled, but part of the process is to translocate ions into non-critical tissues.

Box 2.4 Nutrient deficiency and toxicity symptoms.

Deficiencies of nutrients and excesses of elements lead to changes in tree growth patterns and a range of symptoms develop which may be used for diagnostic purposes. Symptoms may result from interactions and this may be misleading and need to be verified by use of foliage analysis. There is some similarity in symptoms of an element across a number of species but there are also differences. Salt excess will cause scorching of foliage but in initial stages these symptoms may also be caused by other elements. A pattern of effects may be developed for a species and presented in a key: examples of such keys are shown for several species below. The initial key is based on information for *E. camaldulensis, E. grandis, E. pilularis* and *E. tereticornis*. Some notes are added for *E. globulus*, but the presence of waxes may mask affects.

Symptoms in young *E. camaldulensis, E. grandis, E. pilularis* and *E. tereticornis*

A. Symptoms initially appearing in mature leaves

A1. Change in colour primary symptoms

1 Interveinal areas turn pale green/yellow with dark green major veins. As develops, all expanding leaves become pale green/yellow and may be reduced in size until all leaves are affected. There may be spots of anthocyanin with premature leaf shedding. Symptoms move from oldest to youngest leaves.

NITROGEN DEFICIENCY

2 Foliage initially may become dark green. Interveinal purple colouring followed by centre of areas becoming necrotic. Purple anthocyanin colour on underside of leaves then all foliage becomes uniformly purple. Symptoms develop from oldest to youngest leaves and leaf shedding may occur from oldest leaves upwards. Reduced and very variable growth within stands.

PHOSPHORUS DEFICIENCY

A2. Necrosis, scorch or leaf distortions primary symptoms

3 Leaf scorching and necrosis on edges; leaves may be thickened. Remainder of leaf dark green and may be dull in appearance. Dieback of leaf commences at tip and moves back along the leaf blade. Premature leaf fall is common.

SODIUM AND/OR CHLORIDE TOXICITY

4 Scorching of leaf edges affecting all leaf ages. Flagging occurs with one side of tree predominantly affected. Trees located near ocean or spray irrigation systems.

SALT SPRAY

5 Necrosis and scorching initially on older leaves moving to young leaves. Leaves cup, curl or twist. Interveinal tissue turns pale green with the surrounding blade dark green or purple. Necrotic or bleached spots develop. May be associated with increased branching leading to rounded or bushy appearance.

POTASSIUM DEFICIENCY

6 Older leaves very pale green and prematurely shed. Tip scorch develops in mature leaves. Moves from oldest to younger foliage and juvenile foliage develops purple interveinal areas.

MAGNESIUM DEFICIENCY

7 Chlorotic spots in the interveinal areas of mature leaves, only narrow bands alongside the veins remaining green. Purple colour on margins.

MOLYBDENUM DEFICIENCY

B. Symptoms initially appearing in young and developing leaves

B1. Change in colour primary symptoms

8 Young leaves pale green/yellow and progresses to older leaves. In severe cases younger leaves become pale purple with development of necrotic tips.

SULFUR DEFICIENCY

9 Margins of juvenile and expanding leaves pale green. Chlorosis extends between lateral veins towards midribs. Leaves normal in size. Major veins are always flanked with wide margins of green tissue.

MANGANESE DEFICIENCY

B2. Necrosis, scorch or distortions primary symptoms

10 Leaf bleaching and death commencing at top of tree. Symptoms appear very rapidly.

DROUGHT DEATH

11 Younger leaves become reflex and wrinkled with margins turned under. Irregular yellow-green patches appear along the margins and extend inwards towards the midrib. Necrosis then extends inwards from the margins until the leaves die. Terminal shoots on stems and branches die followed by production of shoots. Older leaves may be thicker than normal and show irregular necrotic spots.

CALCIUM DEFICIENCY

12 Interveinal areas become pale green or chlorotic. Leaves become intensely yellow and green tissue only on veins (bleached leaves with narrow green veins). Leaf size often reduced with increasing distance from the stem. Sharp differentiation between chlorotic young and dark green older foliage.

IRON DEFICIENCY

13 Pendulous habit of lateral branches. Expanding leaves twisted, cupped and re-curved and margins irregular. Shoot apex dies back and auxillary buds are shed, develops dwarf multi-stemmed trees without main leader.

COPPER DEFICIENCY

14 Shoot tip growth severely impaired. Shoot tip death followed by repeated re-sprouting and further dieback. Many trees develop multiple leaders and in extreme conditions have rounded, bush-like appearance. Purple pigmentation on young leaves on margins and interveinal areas.

BORON DEFICIENCY

15 Irregular pale green areas near veins or bronzing, xanthocyanin production on leaf edges followed by necrotic areas. Bronzing may occur on some leaves.

BORON TOXICITY

16 Leaves in early stages show mild interveinal chlorosis. Upper surface of the leaf blade show purplish areas distributed between numerous discoloured punctuations. A few circular areas of lighter coloured tissue near the margins with brownish edges. Shortening of internodes leading to rossettes of small narrow yellowish leaves.

ZINC DEFICIENCY

17 Leaf death or scorching on edges. Leaf, branch and/or stem distortion. Occurs in localised patches on trees.

HERBICIDE SPRAY

References: Karschon (1958, 1963a, 1963b); Will (1961a, 1961b); Savory (1962); Hussain and Theagarajan (1966); Truman and Turner (1972); Stewart *et al.* (1981); Morris (1984), Dell *et al.* (1995).

As an example, *P. radiata* is relatively resistant to salt uptake. Foster and Sands (1977) studied seedlings and found salt precipitation in roots, in many cells of the stem and in some needles. In eucalypts a similar process may occur with movement of salt into bark, heartwood and leaves prior to shedding in litterfall. The amount of tissue available for this process is finite and dependent on growth of the tree. When these tissues are saturated, damage to key tissues occurs leading to necrosis and eventually plant death. The available data show that *P. radiata* is an effective rejector of sodium and chloride at relatively high solution concentrations (Sands and Clarke 1977). High chloride levels tended to reduce phosphorus uptake at relatively low solution concentrations so that growth reductions may occur as a result of nutritional interactions.

The pattern of sodium and chloride partitioning is demonstrated within coastally grown *E. grandis* at two different ages (Fig. 2.5). In the young stand (5-years-old), heartwood production was only a minor component of biomass. There was, in absolute quantities, more chloride than sodium. Sodium appears to be relatively low in roots and the two major sinks are in wood and branches. In the 27-year-old stand, chloride was generally higher than sodium. Bark and heartwood are significant sinks for both sodium and chloride and the roots are also a significant sink for chloride. Storage in the stand was about 220 kg/ha for sodium and more than 470 kg/ha for chloride. These results indicate patterns of storage in non-critical tissues but do not provide an indication as to what the maximum storage may be prior to salt damage occurring.

When a comparison of sodium accumulation is made between species (Fig. 2.6), very large differences are apparent. Sodium is lowest in *P. radiata*, and is generally much higher in the eucalypts. Of note is the low quantity of sodium in the *E. tereticornis* roots primarily due to biomass differences, but the highest sodium concentrations tend to be in all tissues in *E. grandis* and higher quantities are accumulated. Sodium is lower in *E. camaldulensis*. While not indicating whether any species is near maximum storage prior to damage, it may be concluded that *P. radiata* rejects or limits sodium uptake. Bark and wood are important for storage of sodium in *E. camaldulensis* and *E. grandis*. In evaluating salt storage mechanisms, both concentration and mass have to be taken into account.

Van der Moezel and Bell (1990) identified three classes of salt effects in eucalypts:

1 rapid stomatal closure and salt exclusion (noted in *E. camaldulensis*)

2 gradual stomatal closure with salt uptake (*E. robusta*)

3 gradual stomatal closure and salt exclusion (*E. microtheca*).

In type 1, stomatal closure precedes salt uptake whereas in type 2, stomatal closure is the result of salt uptake. The ultimate aim in selecting salt tolerant eucalypts for maximum growth and water use in saline soils would be to select trees with high stomatal conductance under saline conditions but which exclude salt. This would then allow plants to survive while transpiring substantial amounts of water, and hence reducing ground water levels (van der Moezel and Bell 1990). In a study comparing relatively salt tolerant *E. leucoxylon* and sensitive *E. globulus*, the greater salt tolerance was associated with higher accumulation of salt in the leaves and stems (uptake by the roots was about the same) (Morris 1981). Salt-accumulating plants were considered to be able to tolerate high foliar salt concentrations if the development of these concentrations was slow enough to enable adaptive processes to occur in the leaf. One part of the adaptive process is thought to be related to the synthesis of proline (Sands and Clarke 1977; Morris 1981).

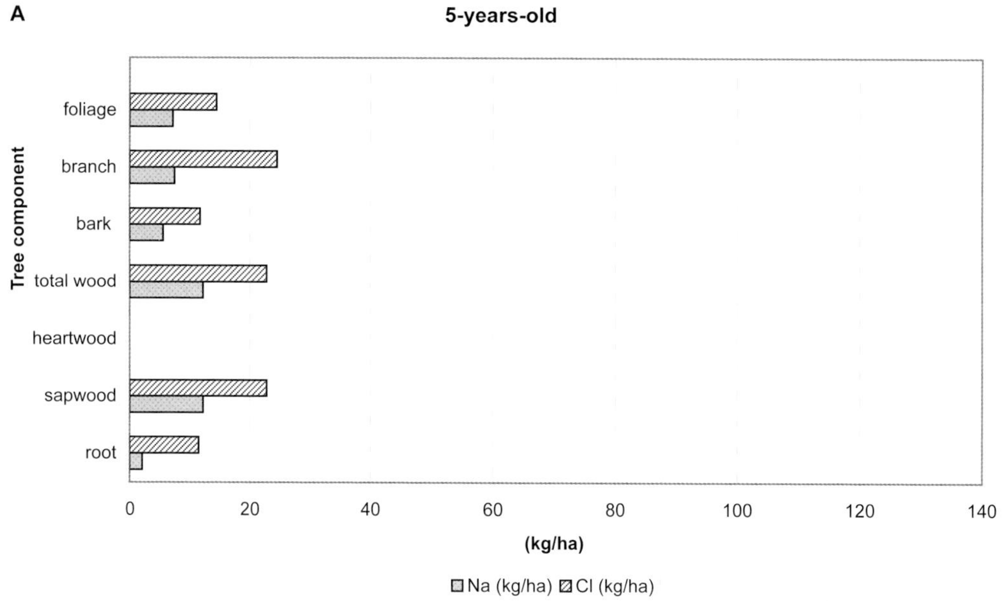

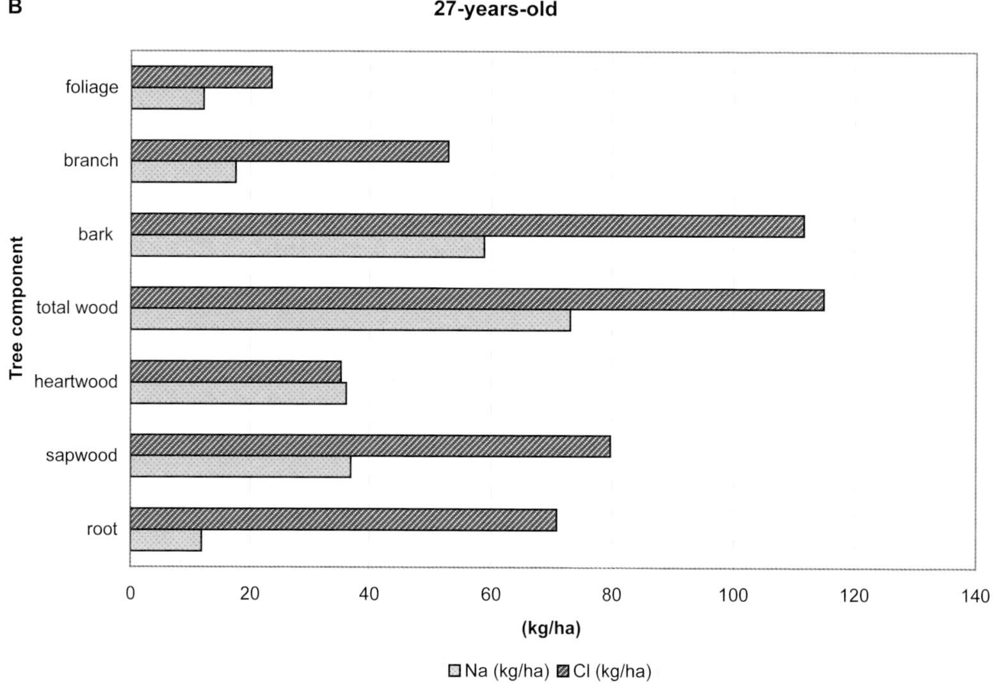

Figure 2.5 Accumulation of sodium and chloride (kg/ha) in different components of (A) a 5-year-old and (B) a 27-year-old *E. grandis* plantation (total wood is the sum of heartwood and softwood) (J. Turner and M.J. Lambert *unpubl. data*).

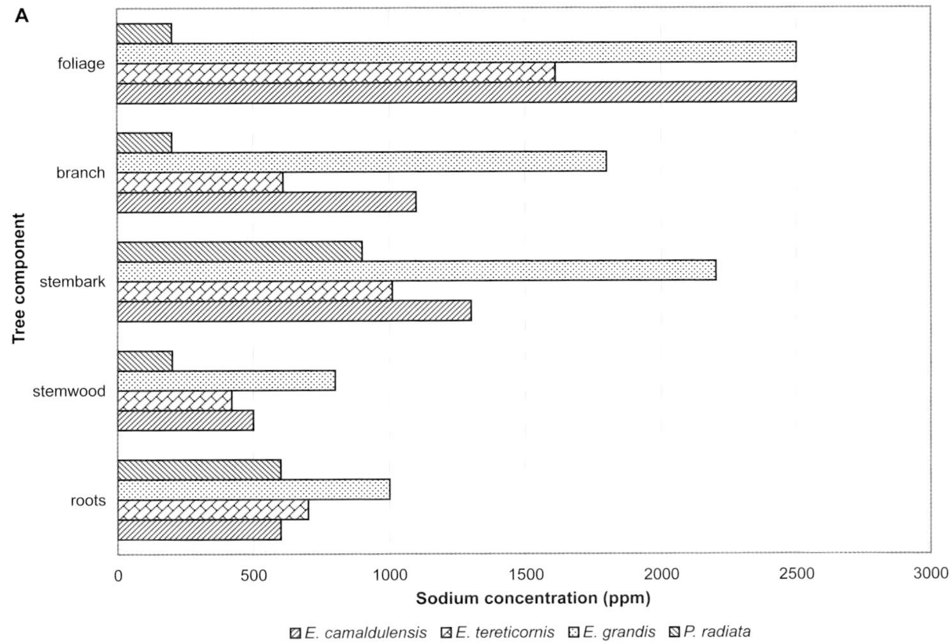

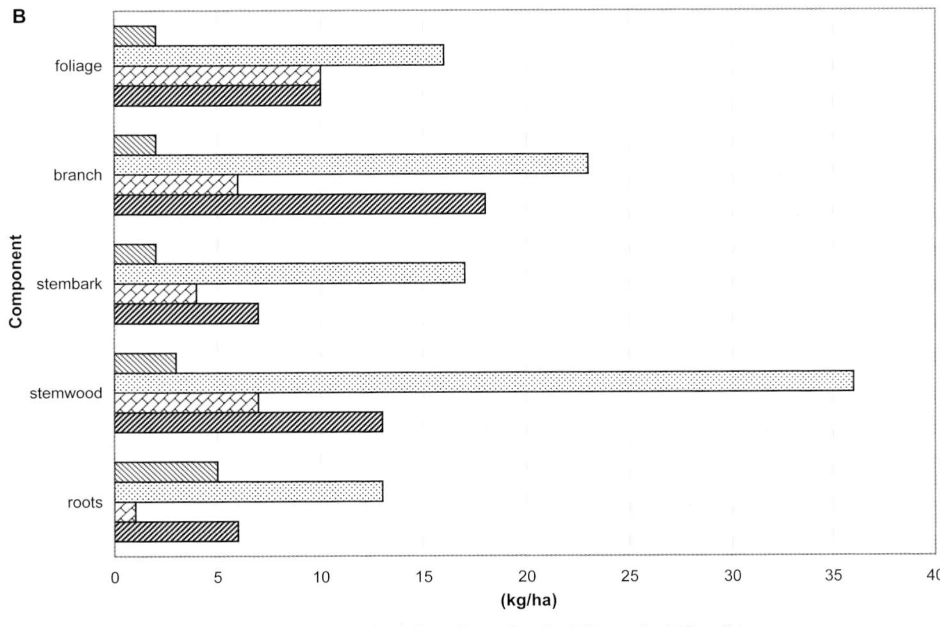

Figure 2.6 Sodium accumulation and compartmentation in four commercial forest plantation species under similar saline conditions. (A) is concentration in tissue, (B) is accumulation in kg/ha. *P. radiata* tends to reject sodium (J. Turner and M.J. Lambert *unpublished data*).

Concluding comments

The assessment of soil salinity in relation to growth of moderately tolerant, deep-rooted, long lived plants has rarely been undertaken. Appropriate standardised methods are required but at this time are unavailable. A range of mechanisms is used by species to reduce the impact of salt and these need to be assessed and understood on an individual species basis.

In assessing commercial plantation species for salt tolerance, there needs to be an understanding of how salt is tolerated. Soil moisture problems are a primary problem and, in addition, there are problems due to toxicities of sodium and chloride and probably other elements.

Growth rate is a critical factor in tolerating salinity as compartmentalisation of the saline elements is a key component of the system. While processes are identified, the critical and limiting levels are not.

CHAPTER 3

SCREENING TREES FOR SALT TOLERANCE

Issues

Tree species vary greatly in their tolerance to both elevated levels of salt and waterlogging. This variation exists between and within species and is also affected by increasing tree maturity. Assessment and screening of trees for tolerance to salt is difficult and there is no consistency in methodology. The purposes of screening for salt tolerance need to be carefully defined and the methodologies need clear presentation.

Selection of species

Establishment of commercial forest crops requires careful selection of the most appropriate and suitable genotypes to meet objectives. Evaluations of commercial tree genotypes for a specific characteristic, such as salt tolerance, need to reflect the end uses of the products. Much work to date has been to screen for species which can survive in soils with elevated salinity or waterlogging, and often in low rainfall areas, primarily for land protection purposes. The requirement in such circumstances is to develop perennial vegetation cover, possibly through the use of direct seeding, but the actual growth rate and production of wood are not key issues. Assessments using survival of young plants as an indicator have covered broad seedlots of a wide number of species growing in saline environments. With more specific objectives, such as lowering of the watertable or production of timber, the actual rates of productivity became a primary interest.

The primary interest in commercial plantations which differ from protection forests, is on the growth rate of selected plants of species with desirable product characteristics. In such circumstances, growth has to be at a minimum level to ensure an economic return (for example, 15 m^3/ha/yr mean annual volume increment over a 25-year rotation) although the required level will vary with the project. The parameter of primary interest is the accumulation of merchantable products over a plantation rotation. Screening for select parameters has been undertaken in a wide number of studies using different techniques. Part of the reason for different screening methods, often using non-comparable methods, is that the primary objectives of the work are quantitatively different (Table 3.1). For example, catchment protection may only require information on survival of species, whereas commercial plantations require information on growth and form of clones. Scale and specificity of required information varies with the objectives.

Analyses of growth or yield of crops in relation to increasing salinity have been standardised (after Maas and Hoffman 1977, Fig. 3.1). This type of system uses proportional decreases in growth and has not been routinely used to evaluate tree species performance. However the

Table 3.1 The basis for selection of species in saline environments for different land management purposes.

	Selection parameter				Selection base			
Purpose	**Seed survival**	**Seedling survival**	**Mature growth**	**Physio-logical processes**	**Species/ provenance**	**Families**	**Hybrids**	**Clones**
Landcare/ catchment protection	**	***	*	–	***	–	–	–
Catchment improvement	*	***	*	–	***	*	–	–
Commercial plantations	–	**	***	–	**	***	***	***
Research (selection)	**	**	**	***	*	*	**	***

Key: *** Very important ** Important * Of interest – Very minor interest

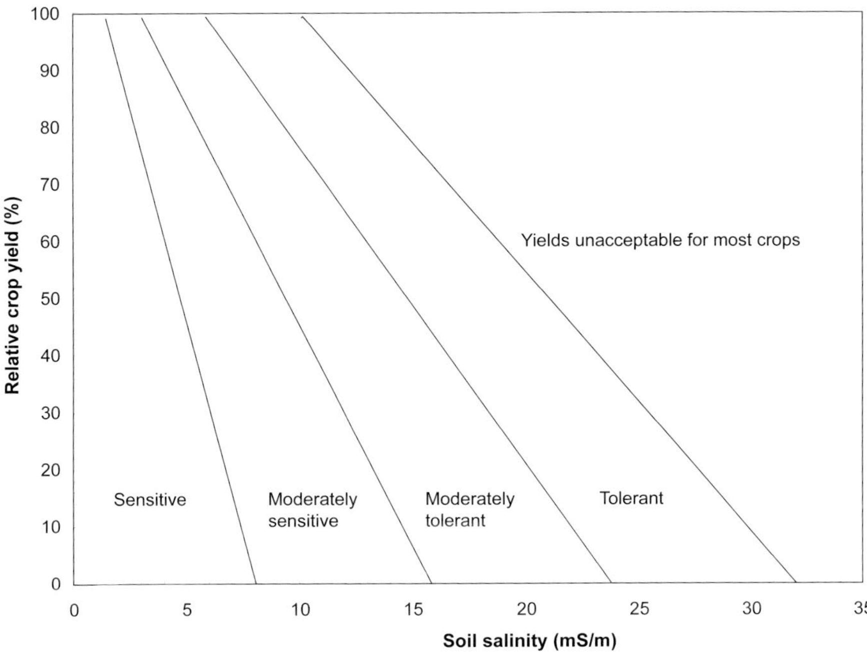

Figure 3.1 Basic graphs from Maas and Hoffman (1977) indicating crop tolerance to salt in relationship to decline in growth.

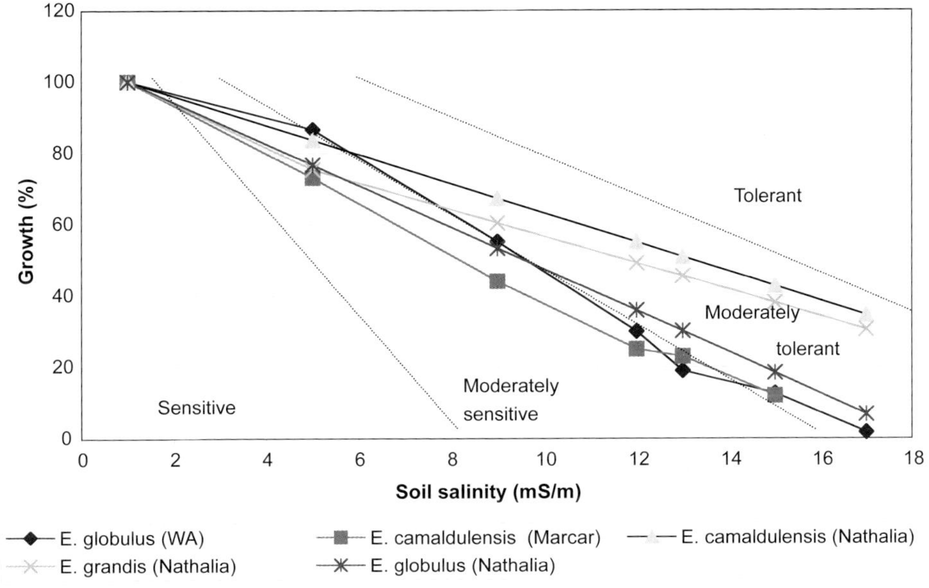

Figure 3.2 Declines in percentage height growth of forest plantation species with increasing soil salinity.

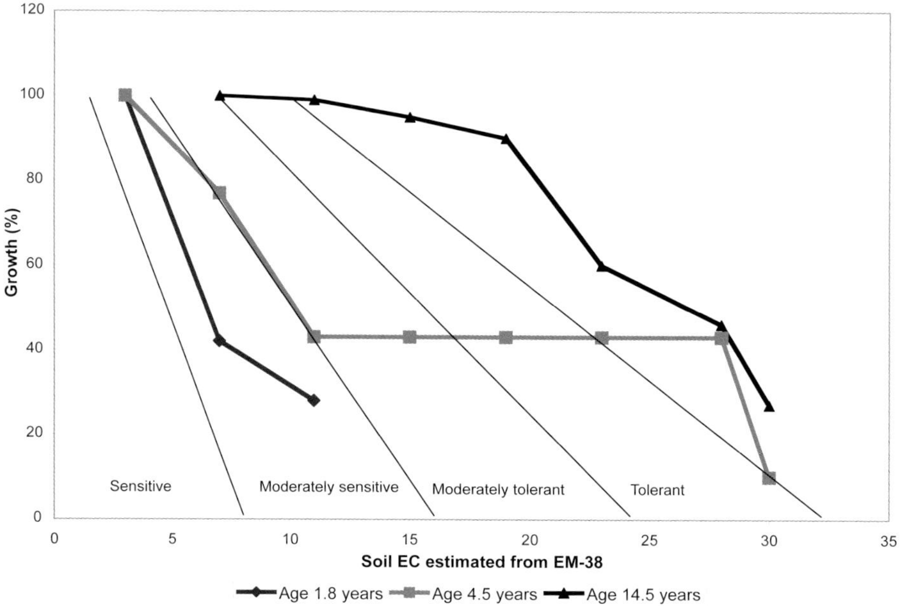

Figure 3.3 Declines in percentage volume growth with increasing salinity, based on EM-38, of three stands of *E. globulus* at different ages (Bennett and George 1995).

system of Maas and Hoffman (1977) is analysed as it provides a standard system for comparison. Preliminary analyses indicate that using relative height growth, *E. globulus* and *E. grandis* gave salinity tolerance comparable with *E. camaldulensis* (Marcar 1995; Hamlet and Morris 1996; P. Shedley *pers. comm.*; Fig. 3.2). This does not mean their absolute growth is similar but that the effect of increasing salt reduces productivity by proportionally similar amounts.

Such a method is valuable to evaluate plants but more assessment data are required for species and analyses of changes in perceived tolerance with increasing plant maturity. In field-grown forest trees, different root configurations and rates of development can affect the tolerance of trees, primarily because trees may be able to utilise zones with less saline water within the soil profile. The salt concentrations in many soils vary with depth. Trees able to develop roots which access these layers could provide an alternative strategy to avoid excessive concentrations of salt. The work of Bennett and George (1995) using EM-38 provides relationships between growth parameters and the presence of salinity (Fig. 3.3). Based on volume growth, there appear to be variations in declines in growth and the age of assessment possibly indicating older stands have alternative mechanisms for handling salt (for example compartmentalisation of salts or deep root development).

In greenhouse studies where plants were treated with different concentrations of saline solutions, Karschon and Zohar (1975) studied three provenances of *E. camaldulensis*. The solutions were sprayed onto the soils or there was partial flooding leading to periodic waterlogging. In terms of absolute growth (mass of roots and shoots), partial waterlogging increased plant mass at low salt concentrations but depressed growth further at high concentrations. Differences in provenances were 20 to 50% for different treatments (Fig. 3.4). When analysed using low irrigated salt concentrations as the control there was little differentiation

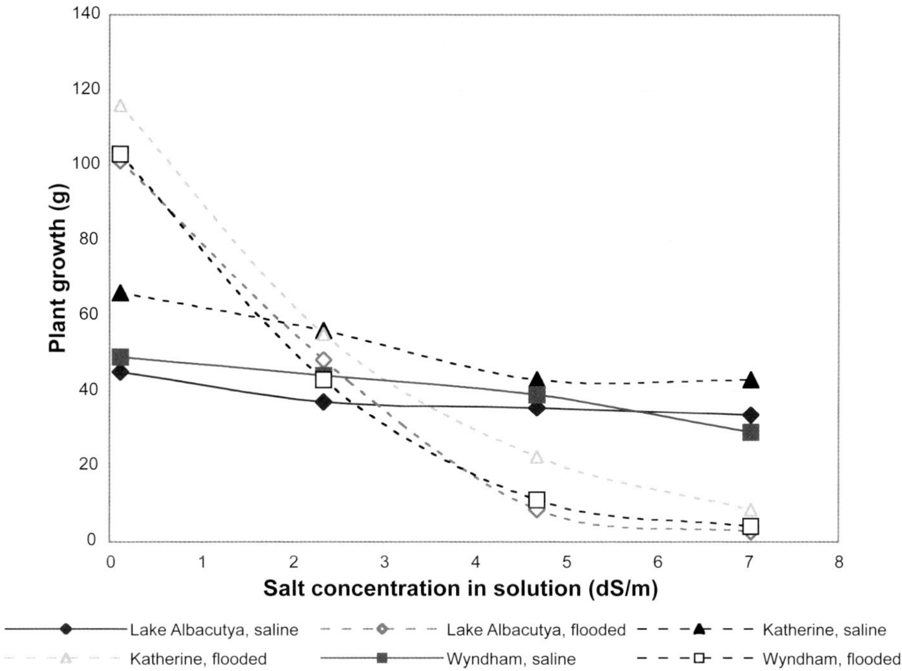

Figure 3.4 Total plant growth (tops and roots) of three provenances of *E. camaldulensis* in solution culture (from Karschon and Zohar 1975).

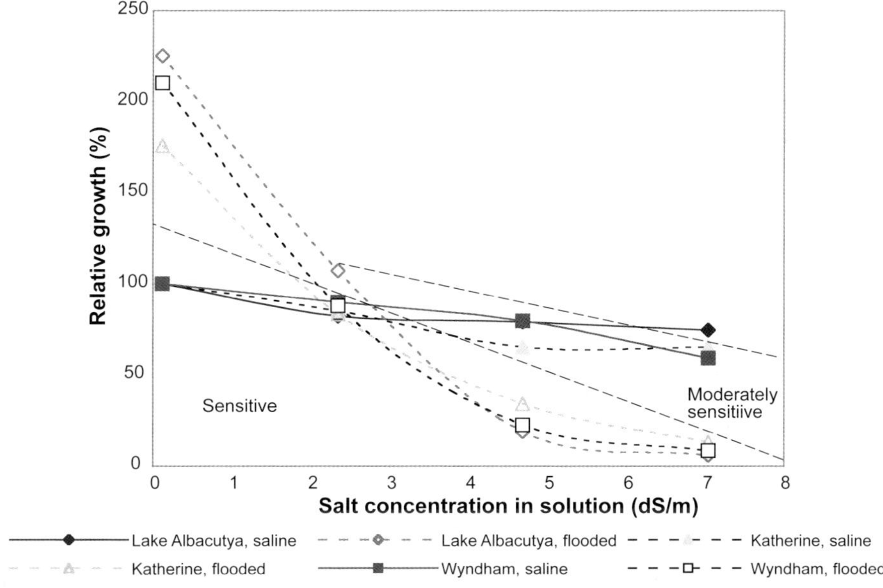

Figure 3.5 Relative growth rate compared with control (low salt, no waterlogging) of three provenances of *E. camaldulensis* in solution culture. The dotted lines represent the Maas and Hoffman (1977) differentiation, indicating the salt concentrations in the study were low, the major impact being from waterlogging.

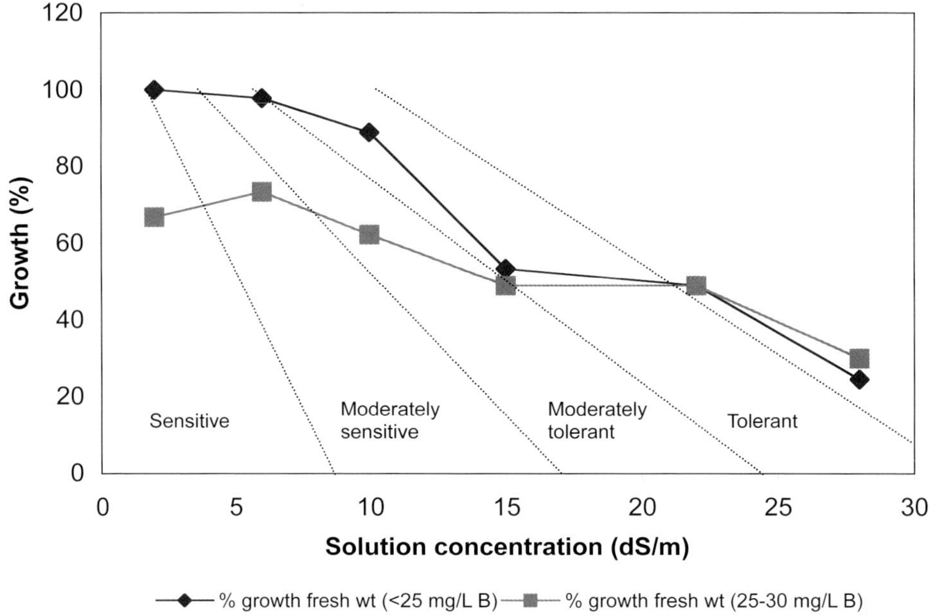

Figure 3.6 Performance of *E. camaldulensis* clone 4544 to increasing salinity, together with high and low boron (Shannon *et al.* 1998). The dotted lines represent the differentiation by Maas and Hoffman (1977).

in response between provenances for salt, while there were significant differences in relative growth at low concentrations. All provenances responded in a similar way when salinity increased (Fig. 3.5). Flooding led to major declines in leaf numbers.

In California, clones of *E. camaldulensis* were studied using lysimeters and solutions dominated by sulfate, some with high boron concentrations (Grattan *et al.* 1997; Shannon *et al.* 1997). The objective was to determine if clones of *E. camaldulensis* were suitable for reducing salinity in drainage water and this required maintenance of high evapotranspiration (ET). In non-stressed environments, the crop coefficient (K_c) for a full cover of eucalypt trees was between 1.2 and 1.5, but the question arose whether salt would reduce this (Stribbe 1975; Shannon 1985). In San Joaquin Valley, *E. camaldulensis* was irrigated with saline drainage water (EC = 10 dS/m and 12 mg B/L) and evapotranspiration was 0.83 (Dong *et al.* 1992) and this lower value was attributed to combined salt and boron. In the study there were increasing concentrations of salt and boron. The most damaged leaves were at lower salt/higher boron and at higher salinity. Uptake of boron was reduced.

Based on the Maas and Hoffman (1977) approach, clone 4544 performed well (more than 50% of maximum growth) in the tolerant range when boron concentrations were low (Fig. 3.6). This performance was better than in other studies but may reflect the composition of the salt solutions, namely they were high in sulfate rather than in chloride. This highlights the effect of salt composition on assessment of tolerance.

Screening of trees for salinity tolerance is a more difficult procedure than screening agricultural or forage crops, primarily due to their longevity, large individual plant size, and physiological changes between juvenile and mature plants. A number of selective evaluations

Table 3.2 Rankings of salt tolerance of commercial species. The rankings of species are relative to each other. In absolute terms, highly tolerant trees species are equivalent to moderately tolerant agricultural species.

	Salt tolerance	Species
INCREASING SALINITY ↓	Moderately salt sensitive	*E. citriodora*[(4)], *E. globulus*[(1,3)], *E. obliqua*[(1)], *E. radiata*[(1)], *E. saligna*[(1)], *E. tereticornis*[(4)], *E. viminalis*[(1)]
	Slightly salt sensitive	*A. mangium*[(6)], *A. mearnsii*[(6)], *A. melanoxylo*[(1,3)], *C. torulosa*[(1)], *E. botryoides*[(1)], *E. cladocalyx*[(3)], *E. elata*[(4)], *E. grandis*[(3)], *E. robusta*[(1)], *E. sideroxylon*[(3)], *Lophostemon confertus*[(1)]
	Slightly salt tolerant	*E. maculata*[(1)]
	Moderately salt tolerant	*A. saligna*[(1,3)], *A. auriculiformis*[(6)], *C. cristata*[(1)], *C. cunninghamiana*[(1,3)], *C. equisetifolia*[(2)], *E. camaldulensis*[(1,3)], *E. citriodora*[(2)], *E. botryoides*[(3)], *E. gomphocephala*[(4)], *E. leucoxylon*[(3)], *E. melliodora*[(3,4)], *E. moluccana*[(3)], *E. robusta*[3,4)], *E. tereticornis*[(3)], *P. halepensis*[(1)]
	Highly salt tolerant	*C. cunninghamiana*[(2)], *C. glauca*[(2)], *C. obesa*[(5)] *E. camaldulensis*[(4)], *E. kondinensis*[(2,4)], *E. occidentalis*[(3)], *E. rudis*[(3)], *P. pinaster*[(1)], *P. radiata*[(1)]
	Tolerant of waterlogged saline soils	*E. camaldulensis*[(3)], *E. ovata*[(3)], *E. robusta*[(3)], *E. rudis*[(3)], *E. tereticornis*[(3)]

[1] Morris (1984), [2] Gill and Abrol (1991), [3] Marcar (1992, 1995). [4] Firman (1968), [5] van der Moezel *et al.* 1988, [6] Marcar and Khanna (1997)

have been developed for species in Australia and are applicable to other areas (for example Table 3.2). The issues that arise in screening are various and involve assessment of genotypes (species, hybrids, provenances, families and clones), survival and growth under increasing levels of salinity, and assessment of changes with plant maturity. Further, there are changes to be considered in the environment with time, specifically changes in levels of salt and composition of salt (either seasonally or in the long term). Such differences will affect the type and method of screening required. Selecting trees to produce a commercial timber crop is more focused on growth performance, and the minimum absolute acceptable growth will be a determinant, rather than the percentage decline in growth.

A number of technical and practical problems arise in relation to screening trees for salt tolerance, and have been reviewed on a number of occasions (for example, Morris 1980). Much of the focus has been on selection of tolerant species and rejection of non-tolerant material, such a process limiting the ability to study physiological processes. The methods used for evaluating salt tolerance for commercial crops have included germination rates (Sands 1981; Clemens *et al.* 1983; Totey *et al.* 1987), growth and yield (Shannon 1985; Allen *et al.* 1994), comparison of growth on saline and non-saline sites (Maas and Hoffman 1977; Sun and Dickinson 1995a, 1995b), tree crown volume response (Marcar *et al.* 1994),

Table 3.3 Potential selection criteria for assessing salt tolerance (Tal 1985).

Growth	
Germination percentage Emergence percentage Seedling survival	Useful only for crops that are equally resistant during germination and later stages
Rooting	Configuration and rate of root development
Ions	
Accumulation	Indicates tolerance in salt accumulators
Exclusion	Indicates tolerance in salt excluders
Membranes	
Leakage of electrolytes Abnormal plasmolysis	Indicates sensitivity
Vital staining	May indicate disturbed/undisturbed metabolism
Chlorophyll fluorescence	Changes in parameters may be used for screening for salt tolerance

ion concentrations in various tissues (Townsend 1989), changes in water status, stomatal conductance, and/or net photosynthesis (Allen *et al.* 1994). Tal (1985) outlined physiological parameters potentially of value for salt screening (Table 3.3), however none of these have been used consistently for forest trees.

Plant development stages at which screening has been proposed are individual cells (Hasegawa *et al.* 1986), tissue segments (Morabito *et al.* 1994), young plants in solution culture, sand or soil culture (Morris 1980; Clemens *et al.* 1983; Shannon *et al.* 1998), and younger and older plants in the field (Firmin 1968; Mathur and Sharma 1984; Pepper and Craig 1986; Morris *et al.* 1994; Greenwood *et al.* 1995; Sun and Dickinson 1995a, 1995b). Screening of tissue segments and *in vitro* assessments consider physiological tolerance by growing tissues, and are of limited relevance to forest tree species as many of the mechanisms of tolerance in trees appear to function by salt avoidance. While all these studies test effects of salinity, they analyse differing components of tolerance.

Screening for 'salinity' alone accepts that the trees are growing in high salt environments. However when undertaking a tree selection program, the components of tolerance need to be determined and, as such, screening will need to be more specific and detailed. Types of screening undertaken for testing tree salinity include glasshouse and field testing and require validation using longer term field trials.

Laboratory assessments

Tissue assessment

Selected tissues, segments and seeds have been subjected to saline levels to ascertain survival and physiological changes. In forest trees, this has not been related to productivity but has been of some interest in ecological relationships (Clemens *et al.* 1983; Morabito *et al.* 1994). The periods of time of assessments tend to be days or weeks.

Glasshouse testing

Testing of material in the glasshouse requires use of relatively small plants and considers shorter term growth in a specific growth medium. Any such medium can utilise seedlings, cuttings and tissue culture-derived plants, and evaluate the plants using a range of morphological and physiological testing techniques (for example Farrell *et al.* 1996a, 1996b). That is, in a testing system, plants may be grown in soil, sand, fixed solution or flowing solution or an inert medium and with this, seedlings, cuttings or other plantlets can be grown. Within the matrix, the period of study will add further complexity and variation and hence the type of study needs to be carefully defined. Comparability between trials by different researchers is difficult. The different forms of evaluation are discussed.

Single fixed concentrations (of applied solution)

Solution culture has often been preferred rather than soil, since treatments are not affected by cation exchange problems and fluctuations in moisture and salt content are not factors (Richards 1969). Use of solution culture requires careful design and use of the solutions, since even minor changes in composition, such as changes in ratios of sulfate and chloride or the form of nitrogen supplied, can result in significant changes in growth and performance (Kafkafi 1984). Single fixed concentrations use a single growth solution in which plants are grown and test specific responses, although within an experiment there is a range of treatments. Truman and Lambert (1978) used a base solution of calcium, magnesium, phosphorus, nitrogen and trace elements. Separate levels of salts were added and the ratio of sodium to potassium was maintained at 50:1. Overall, sodium chloride was increased from 2.5 mM to 460 mM giving 10 separate treatments (that is sodium concentrations in the separate treatments were 2.5, 5.0, 10.0, 20.0, 40.0, 80.0, 160.0, 260.0, 360.0 and 460.0 mM with K as 0.02 of each treatment). The technique allows the testing of relative survival rates between genotypes or the rapid screening of a large number of species. The selection of solution composition and concentration, aeration, recycling and replacement times of solutions are all critical issues. Shannon *et al.* (1997) used single solution but all cations and anions (other than nitrogen and phosphorus) were increased and the overall EC increased.

Increasing concentrations over time (of applied solution)

In such studies (Fig. 3.7), there are increasing levels of salinity which are applied over time and survival and growth are monitored (Blake 1981; Felker *et al.* 1981; Bell and van der Moezel 1988; Rhodes and Felker 1988; Marcar and Termaat 1990; Oddie *et al.* 1996). The point at which 50% survival or growth occurs can be determined as an index of performance (for example Morris 1984). This method has been used to test a wide range of species including many tree species and genera (Townsend 1980). The increases in concentrations with time allow plants to adjust, but there is a simultaneous increase in maturity and changes in environments which can confound results. Further, the period of time in solution tends to be short. Morris (1984) carried out glasshouse screening using increasing solution concentrations on 6- to 12-month-old plants. He determined salt concentrations causing 50% mortality and 50% reduction in height growth together with an overall rating of tolerance. He noted that salt stress symptoms included initial wilting of growing tips progressing to complete wilting and tip death, though in a few species slight wilting and low concentrations were followed by recovery, presumably indicating physiological adjustments. The leaves of salt-treated plants usually became duller or lighter in colour than controls.

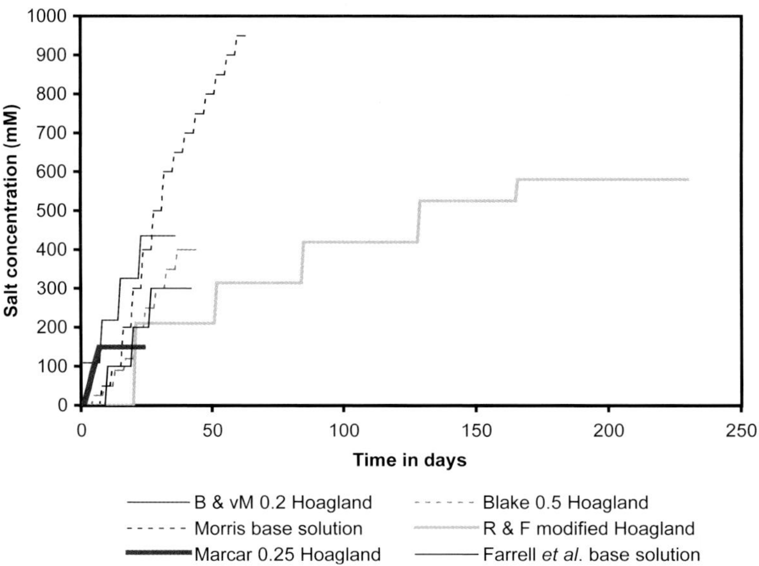

Figure 3.7 Comparison between solutions and treatment times from a selection of studies using increasing quantities of applied salts. Solution concentrations have been converted to approximate mM for comparative purposes. The day of application and the strengths or relative strengths of salt are shown.

Serial concentrations (of applied solution)

Plants are grown for lengthy periods in a series of solutions which will increase in salt concentrations. The solutions are specifically designed and may vary in a number of factors (for example, to maintain osmotic potential) whereas the 'increasing concentration trials' use a standard solution with direct additions of salt. While there may be a period of adjustment, particularly osmotic adjustment at higher concentrations, the method allows longer-term analyses to be undertaken to assess relationships between salinity and plant growth decline. The design of such cultures may include crossovers (that is, maintenance of equivalent total concentrations of two ions but exchanging elements from high to low) to maintain osmotic potential (see Table 3.4). In that study, potassium was freely taken up and translocated to the shoots, the levels in the shoots being higher than those in the roots. However, the levels of potassium in both shoots and roots were significantly less in solutions in which sulfate predominated over chloride. Uptake and translocation of sodium was restricted, the ratio of sodium (shoots) to sodium (roots) being less than unity. The use of increasing gradients of salinity, rather than a set of fixed concentrations applied over a long period, avoids salt shock (dehydration) resulting from sudden application of a high salt concentration and enables assessment of a large number of species in a small area in a short period.

Serial concentrations (soil or sand)

Such a process involves a similar principle to the one above but other factors such as waterlogging may be addressed by using solutions applied to washed sand or soil (Clemens *et al.* 1983). Shannon *et al.* (1998) used a similar technique but with sand in very large lysimeters, and addressed increasing salinity and boron with relatively large trees. Such a technique

Table 3.4 Crossover concentrations in Experiment No. 1 used by Truman and Lambert (1978) and the resultant potassium concentrations in shoots and roots of *Auracaria heterophylla* to demonstrate effects.

	Solution K/Na (mM)					
Solution SO_4/Cl (mM)	**2.1/0.1**	**1.7/0.5**	**1.3/0.9**	**0.9/1.3**	**0.5/1.7**	**0.1/2.1**
(a) Potassium, shoots (μmol/g dry wt)						
1.25/0.1	588	562	485	547	493	353
0.85/0.9	581	569	558	548	478	323
0.45/1.7	607	595	565	577	550	346
0.05/2.5	610	611	599	558	535	404
Mean	596	584	552	557	514	336
(b) Potassium, roots (μmol/g dry wt)						
1.25/0.1	452	419	384	414	364	226
0.85/0.9	440	410	374	392	362	204
0.45/1.7	524	450	417	438	397	215
0.05/2.5	510	435	402	411	399	219
Mean	481	428	394	414	380	216

allowed analysis of conditions comparable with the field conditions plus extremes of salinity on trees with sizes representative of field plantings.

Soil studies

Soil studies include glasshouse trials using plants grown in various types of soils possibly with amendment and they especially allow testing of interactions to be undertaken with various degrees of waterlogging (Farrell *et al.* 1996a, 1996b). In relation to salinity, these are often more difficult to control because of variability and other interactions, however the effect of soil texture can be included. Modified soil salinity zone effects have been developed and used for grapevines indicating advantages for morphological studies, and these may be applicable to trees (Sykes 1985).

Field trials

'Field trials' generally involve establishment of trees in environments typical of routine plantings. While desirably they would involve replicated experiments with an understanding of the variable (especially salinity levels), they often involve unreplicated plantings and possibly demonstration areas with only limited information on the soil conditions. Such trials are dependent on available rainfall or ground water for growth.

Trials in saline areas

Establishment of trials in known saline conditions allows the assessment of relative (between genotypes) survival and growth to be undertaken. It is, however, difficult to obtain a true control which is desirable for assessment of impacts of salt on growth, while the variability of soil salinity makes within-trial variability difficult to interpret (Zohar 1982). Techniques such as assessment with EM-38 provide some data (Bennett and George 1995). It is possible to utilise sufficiently extensive data, especially with multiple well-designed trials, to establish relationships between growth and salinity. Field trials have been established in a large number of countries using both exotics and local species but standard designs and methodologies have, as yet, not been developed (Sheikh 1974; Malik and Sheikh 1983; Hussain and Gul 1992; Hafeez 1993).

While not specifically studying selection for salt tolerance, the use of tree ring widths has indicated effects of changes in soil water salinity over long periods of time and may in some cases provide historical growth data for some field grown trees in saline conditions (Yanosky *et al.* 1995; Yanosky and Kappel 1997). The activity factor in trees together with changes in environmental conditions over time is a factor that requires attention.

Irrigation with saline water

Irrigation water may be applied to supplement tree growth but also provide a method of applying salts for assessing tree response. Irrigation with saline water compared with non-saline water should provide conditions which will lead to an understanding of salinity levels in soil solutions in the root zone. The trial design has to be explicit and, preferably, different levels of salt need to be applied. Such trials are relatively large scale but have been used to provide valuable longer term data on the relationships between salinity and forest tree response (Stewart *et al.* 1988).

Comparison of results from glasshouse and field trials

Results obtained from different forms of assessment of response to salt need to be consistent to provide comparability. Unfortunately, there has been little direct work done on comparisons of performance of seedling material and field grown trees. Analyses which compare increasing concentrations of salt in greenhouse studies with the same genetic material in field trials and at 50% cut off in growth rates would be optimal, but generally are unavailable.

Few if any trials have been established which analyse mechanisms associated with salt tolerance but such designs are of critical importance. There have been a number of field trials covering a large number of species (for example Firmin 1968; Pepper and Craig 1986), others in solution culture (for example Blake 1981; Morris 1981) plus others with smaller numbers of species. In evaluating results from each of these types of trials there is a problem comparing levels of tolerance, since the assessments are generally quite subjective and the results vary between researchers. There is no standard or consistent definition of tolerance levels in either field or glasshouse studies so the individual author's assessments have been used where the same species has been tested in the field and in the greenhouse (in all cases by different researchers) and a conclusion drawn. Several studies (including for example, *E. microcorys*, *E. camaldulensis* and *E. globulus*) have produced very variable results which are, in part, a result of genetic variability.

Recognising the differences in methodologies and evaluation and by generalising the solution culture and field studies, a comparison of results and assessments has been made to assess whether the results from glasshouse studies provide an indication of those found in the field (Table 3.5). Field and glasshouse (seedling) evaluations apparently had both been undertaken on 63 tree species. Twenty-one species provided similar results, in that the greenhouse evaluation indicated a comparable level of salt tolerance as in the field trials, while for the remainder of the species, the greenhouse evaluation over- or under-estimated salt

Table 3.5 Comparison of reported tolerance of eucalypts to salinity between solution culture and field grown trees. Hatched areas indicate a match between glasshouse and field testing.

		Glasshouse tolerance				
		Very high	**High**	**Moderately high**	**Low**	**Very low**
FIELD TOLERANCE	Very high	*E. camaldulensis*	*E. microtheca* *E. moluccana* *E. sargentii* *C. cunninghamiana*	*E. occidentalis* *E. spathulata*	*E. kondinensis* *E. obliqua*	
	High	*E. gomphocephala*	*E. dundasii* *E. incrassata* *E. landsdowneana* *E. largifiorens* *C. equisetifolia* *C. obesa*	*E. botryiodes* *E. cladocalyx* *E. pellita*	*E. globulus* *E. intermedia* *E. microcarpa*	*E. lesoueffie* *E. salubris*
	Moderate	*E. brockwayi* *E. wandoo* *E. woodwardii*	*E. intertexta* *C. cristata* *C. glauca* *P. radiata*	*E. astringens* *E. campaspae* *E. laucoxylon* *E. rudis* *E. torquata* *E. urnigera*	*E. alba* *E. saligna*	*E. robusta*
	Low	*E. argopholia* *E. tereticornis*	*Acacia saligna* *E. maculata*	*E. citriodora* *E. grandis* *E. revertiana*	*E. aggregata* *E. albans* *E. cinerea* *E. diversifolia* *E. globoidea* *E. ovata* *Lophostemon confertus*	*E. globulus* *E. pilularis* *E. radiata*
	Very low				*E. albans* *E. microcarpa* *E. muellerana* *E. sieberi* *E. viminalis*	

Aswathappa and Bachelard (1986); Bacon *et al.* (1993); Bidner-BarHava and Ramati (1967); Blake (1981); Firmin (1968); Fox *et al.* (1990); Greenwood *et al.* (1994); Hoy *et al.* (1994); Ladiges and Kelso (1977); Marcar *et al.* (1991a); Marcar *et al.* (1995); Mathur and Sharma (1984); Morris (1981); Morris *et al.* (1994); Parsons (1968); Pepper and Craig (1986); Pettit and Froend (1992); Prat and Riyad (1990); Sun and Dickinson (1993); Sun and Dickinson (1995b), J. Turner and M.J. Lambert (pers. comm.); van der Moezel and Bell (1990); Zohar (1982); Butcher (1983). Note: *E. microcorys* has been reported as having both very low and moderately high salinity tolerance, *E. globulus* from low to high, and *E. camaldulensis* is listed as very high by general assessment but is variable.

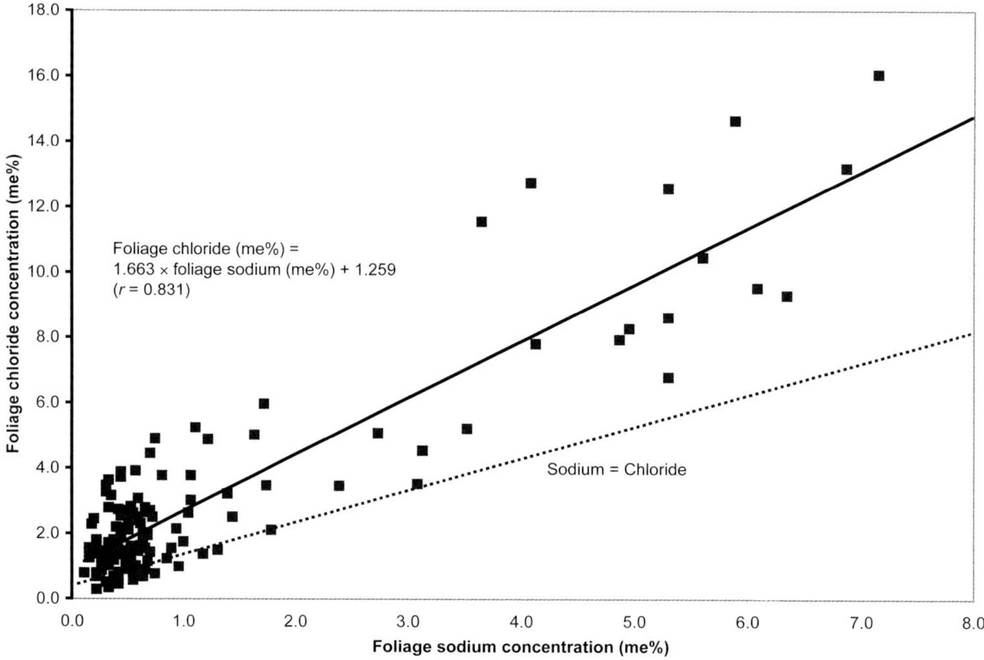

Figure 3.8 Relationship between foliage concentrations of sodium and chloride in field grown *P. radiata* (Turner 1972).

tolerance in the field. That is, in the greenhouse 22 species were estimated as having a lower tolerance than demonstrated in the field, while 20 were over estimated. While only a coarse evaluation of performance, it probably is a reflection that attributes assessed in the field were different to those in the greenhouse. It may be concluded that the greenhouse methodology requires more calibration as a predictive system although it is currently important for physiological evaluation.

One factor affecting differences in field trials is rate of growth. Rapidly growing trees are more tolerant of soil salinity than slow growing trees. While this becomes a circular argument since the assessment of salt tolerance is undertaken by assessing growth, basically in the case of moderately tolerant species using avoidance mechanisms, growth rate is a critical factor. For example, Sun and Dickinson (1993) found that responses to salinity of plant leaf number and height were indicative of tolerance to salt. Species with an ability to maintain a relatively large number of leaves and relatively large height growth when subjected to salinity appeared likely to be more tolerant to salt.

Evaluation of salt in tissues

Determination of elemental concentrations in plant tissue is one method of assessing response to increasing salinity. Basically, the method is based on developing empirical relationships between foliage element concentrations and symptoms or changes in growth, and is generally expressed on the basis of dry weight of tissue and determined in photosynthetic tissues (Rogers 1985). However some studies have utilised salt concentrations in tissue water

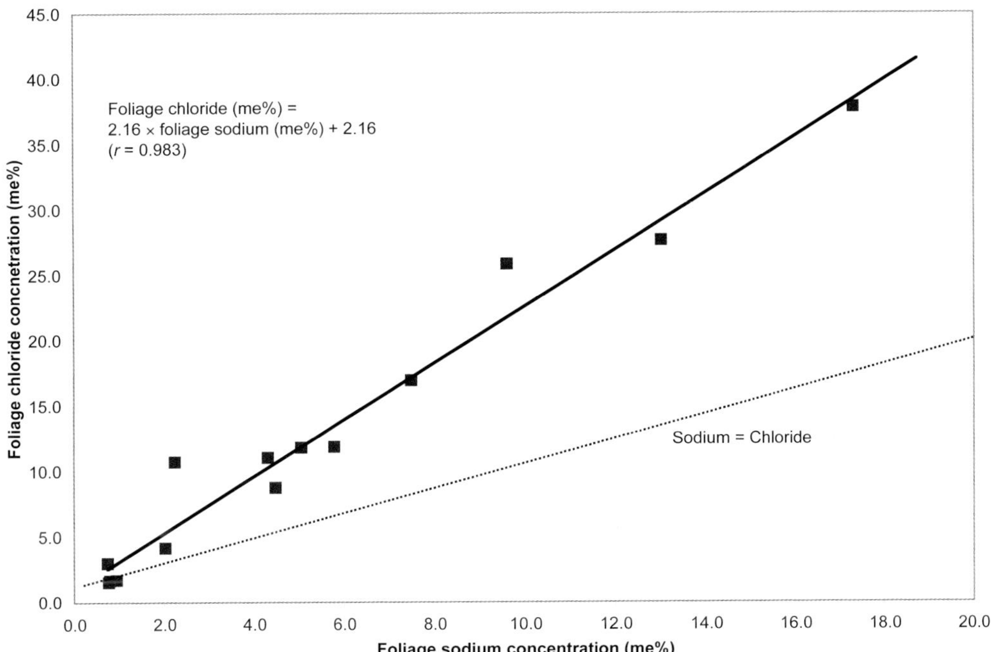

Figure 3.9 Relationship between foliage concentrations of sodium and chloride in field grown *P. patula* (Turner 1972).

as an index (Walker *et al.* 1993). Regardless of the method used, there needs to be calibration of results relating concentrations of elements to plant health or growth parameters. Assessments show that there are differences within and between eucalypt species but no definitive relationships have been developed with salt tolerance. This is partly due to different methods used by plants to handle the high levels of salt in the environment, and these alternative mechanisms need further analysis.

Studies involving a significant number of chemical analyses over a range of sites have included sodium and chloride balances in tissues of tree species. Alternative models for sodium and chloride in foliage may be considered including:

1 Sodium and chloride are taken up and translocated to the foliage in a largely balanced form (that is Na = Cl). At excess levels, salt would be a problem.

2 Sodium is taken up and/or translocated to foliage generally in excess of chloride (that is Na > Cl). Sodium would be identified as a problem.

3 Chloride is taken up and/or translocated to foliage greatly in excess of sodium (that is Cl > Na). Chloride would be identified as a problem.

4 In known high salt environments, sodium and chloride foliage concentrations are low in which case the plants are effective excluders or limit translocation to foliage.

In radiata pine which is a relatively salt tolerant tree species, there is a pattern of reducing or modifying uptake of sodium to foliage but accumulating chloride in relatively higher concentrations (Fig. 3.8). While correlations can be found between sodium and chloride, chloride

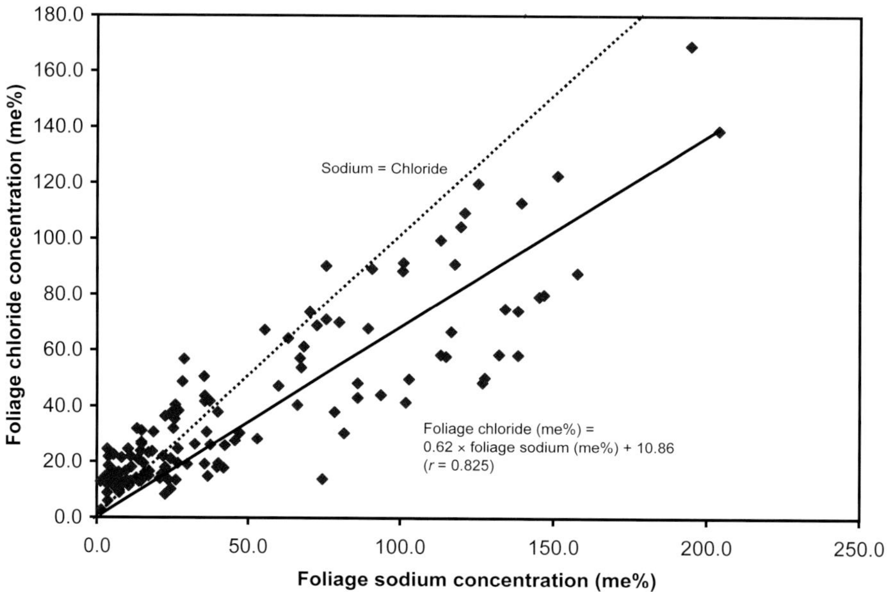

Figure 3.10 Relationship between foliage concentrations of sodium and chloride in field grown *Araucaria heterophylla* (Truman and Lambert 1978).

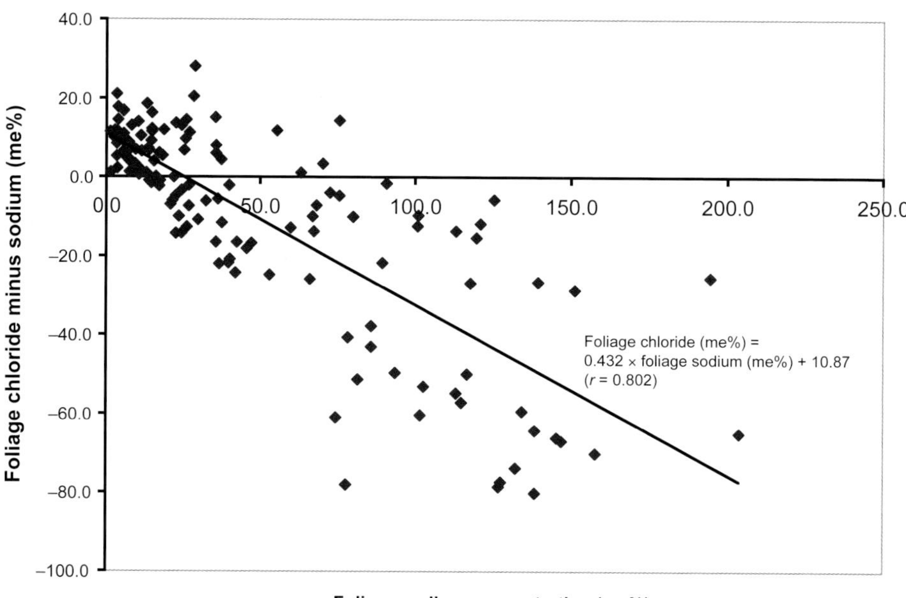

Figure 3.11 Relationship between foliage concentrations of sodium and excess chloride (chloride minus sodium in me%) in *Araucaria heterophylla* (Truman and Lambert 1978).

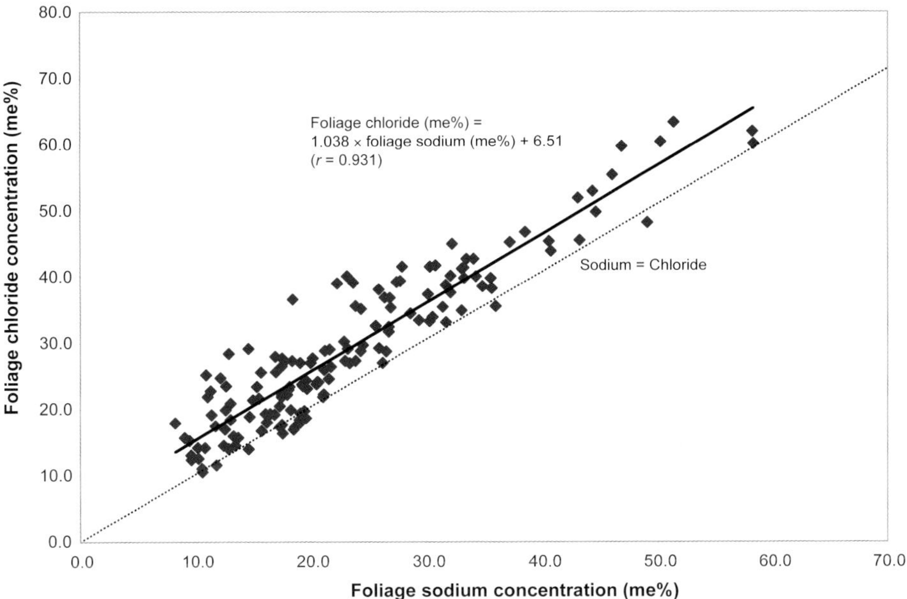

Figure 3.12 Relationship between foliage concentrations of sodium and chloride in field grown *E. pilularis* (Turner 1972).

is usually 4 to 5 times higher (in equivalents) than sodium. In this species, damage assessment and monitoring should be related to chloride concentrations, although sodium has been used as an index of damage as this is the element more routinely assessed (Woods 1955; Turner and Kelly 1973; Sands and Clarke 1977). Based on the above models, chloride is expected to be the main problem. In *P. patula*, a species readily demonstrating salt damage symptoms, sodium and chloride are highly correlated (Fig. 3.9), however chloride is accumulated at about twice the level of sodium and damage appears to be due to chloride, although it is not easy to differentiate. In similar conditions, there are much higher foliar salt concentrations in *P. patula* than in *P. radiata*, and *P. patula* does not appear to avoid sodium to the same extent. These patterns indicate possible differential accumulation methods but they are not directly related to plant growth.

Norfolk Island Pine (*Araucaria heterophylla*) is a species which is very tolerant of salt. It has been demonstrated that it can be grown in solutions with high salt concentrations without apparent damage. However salt damage occurs when the waxy leaf coating is removed or damaged by surfactants allowing the ingress of salt through the foliage. There is not a direct balance between sodium and chloride (Fig. 3.10) but excess chloride shows a pattern (Fig. 3.11). At low concentrations, chloride tends to be much higher than sodium but with increases in concentration, sodium is taken up in excess of chloride, and in this situation it is considered that sodium is the damaging ion. While this is not related to uptake mechanisms, it demonstrates the problems of foliage interpretation. Based on the models proposed, sodium would be the main problem element for this species.

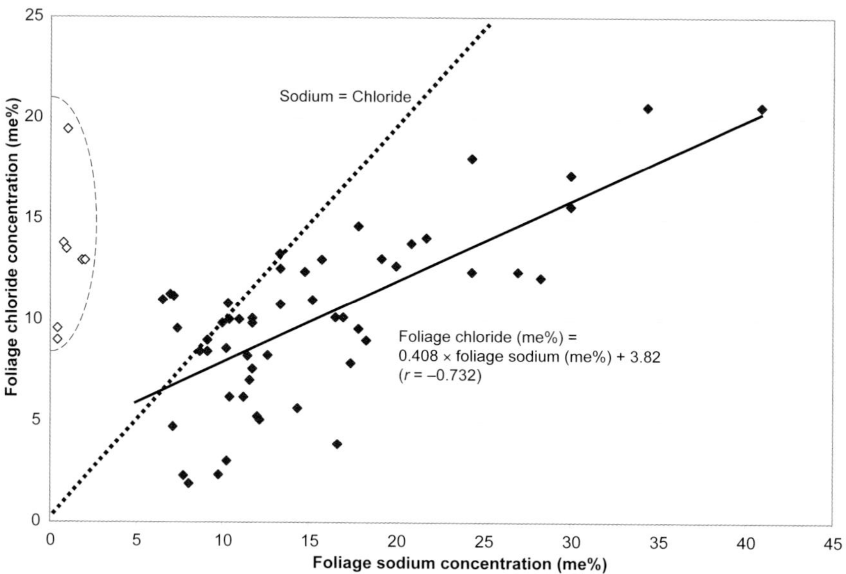

Figure 3.13 Relationship between foliage concentrations of sodium and chloride in *E. camaldulensis*. The isolated figures (in dotted line) were from an effluent trial which was omitted from the regression.

In blackbutt (*E. pilularis)*, a species with low salt tolerance, high uptake of sodium and chloride is almost in close balance, that is sodium and chloride are taken up in equivalent quantities (Fig. 3.12) and any damage is due to salt rather than either ion.

E. camaldulensis is relatively tolerant of salt, and sodium is accumulated in excess of chloride possibly indicating sodium may be used as an indicator in that species (Fig. 3.13) and that any damage may arise from that ion. Variation in tolerance was found to be large in the red gums (*E. camaldulensis, E. tereticornis, E. rudis*) and this was difficult to chemically interpret at the species level (Thomson *et al.* 1987). Primary factors found to be associated with higher tolerance within and between species appeared to be maintenance of lower foliar levels of salt concentrations, especially chloride and magnesium. Only a broad relationship between sodium and chloride was found in the evaluation of foliage element concentrations in the relatively salt tolerant eucalypt, *E. camaldulensis.* The estimate of chloride excess was comparable (but at lower concentrations than in Fig. 3.14). Analysis of the different patterns indicates that foliage analyses of each species need to be calibrated and interpreted individually, and that the cause and degree of toxicity problems vary.

Species of *Casuarina* showed different mechanisms when comparisons were carried out between species (Clemens *et al.* 1983). For *C. equiistifolia* and *C. cristata,* there was storage of salt in roots, *C. cunninghamiana* stored salt in stems, while in *C. inophloia* salt was transferred straight to the tops, demonstrating effective compartmentalisation in some species.

Interpretation of foliage chemical composition requires consideration of net effect of root uptake and transport, together with intake through foliage. An approximate direct balance of sodium and chloride is often associated with intake through foliage (Cassidy 1968). Imbalances are either avoidance or compartmentation but the element in highest concentration may be assumed to cause damage. Several different models may be considered.

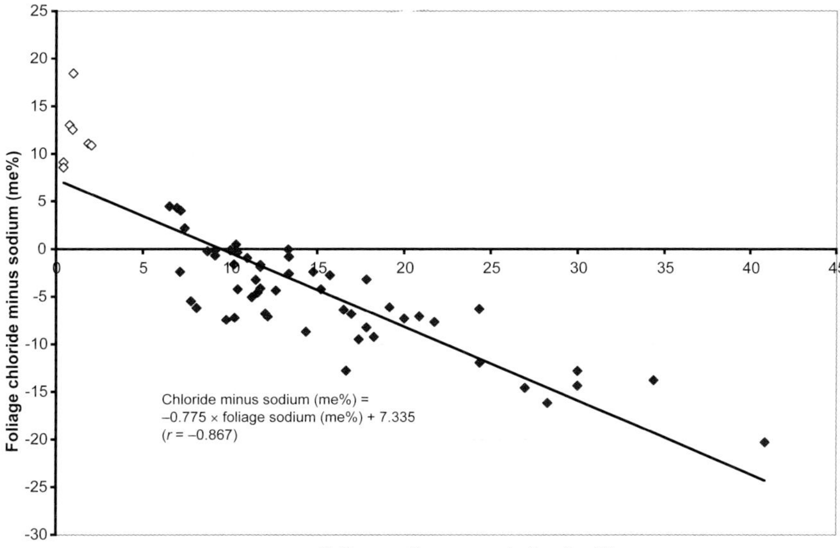

Figure 3.14 Relationship between foliage concentrations of sodium and excess chloride (chloride minus sodium) in *E. camaldulensis*.

Sodium and chloride concentrations in the bark, sapwood and heartwood components of trees provide indices of salt compartmentation by trees. For such an evaluation, the trees need to be large enough to develop heartwood and large quantities of bark. A range of patterns emerge (Table 3.6). For example, *E. maculata* appears to store chloride in heartwood and bark. *E. cypellocarpa* is utilising bark for compartmentalisation of both sodium and chloride. The pattern for sodium does not give any indication of what the pattern will be for chloride and both need to be evaluated separately. A species such as *Syncarpia glomulifera* is of interest as little or no compartmentation has occurred and it generally has low salt tolerance. Such a method of assessing a range of tissues provides an estimate of alternative tolerance mechanisms. Foliage analysis is valuable but needs calibration for each species.

Salt tolerance and waterlogging

Waterlogging is often associated with saline soils because:

- Most of the root zone is already saturated or moist either due to rising watertables or seepage from perched watertables higher in the landscape, and/or
- Infiltration and permeability are low due to soil sodicity.

In many areas, waterlogging by itself is a major problem and is often more widespread than salinity. The principal plant stress from waterlogging is an oxygen deficit. Periods of inundation or flooding will cause added stress. Declines in oxygen can occur rapidly during flooding of soils. After one day of flooding, soil oxygen was found to decline from 5 mg/L to less than 1 mg/L (Davison and Tay 1985). Very few field trials have been carried out in relation to waterlogging and such trials are difficult to design (Marcar *et al.* 1991a). Effects of waterlogging have been studied in a range of eucalypt species in generally low salt environments and the findings indicated that there were very large differences both between and within species (Parsons 1968; Ladiges and Kelso 1977).

Table 3.6 Concentrations of sodium and chloride in bark, sapwood and heartwood of selected commercial tree species in NSW (Lambert 1981). Data in bold are for tissues where sodium or chloride have accumulated.

		Heartwood		Sapwood		Bark	
		Sodium	Chloride	Sodium	Chloride	Sodium	Chloride
Species				(me%)			
E. gummifera	Corymbia	0.10	0.14	1.11	0.54	2.96	4.20
E. intermedia	Corymbia	0.10	0.07	0.61	0.99	3.44	7.34
E. maculata	Corymbia	0.70	**3.20**	0.67	1.89	2.48	**8.85**
E. fastigata	Monocalyptus	0.26	0.14	1.22	1.04	3.91	5.24
E. obliqua	Monocalyptus	0.41	0.04	**3.52**	**5.01**	8.11	7.90
E. pilularis	Monocalyptus	0.39	0.04	0.70	0.31	2.98	4.37
E. radiata	Monocalyptus	0.24	**2.08**	1.33	1.55	4.43	**18.61**
E. camaldulensis	Symphyomyrtus	0.17	0.90	1.56	2.56	4.61	6.92
E. cypellocarpa	Symphyomyrtus	0.26	0.04	2.87	4.84	**8.04**	**15.72**
E. grandis	Symphyomyrtus	0.74	0.56	1.45	1.47	6.04	7.60
Lophostemon confertus		**2.04**	**10.90**	0.65	3.04	0.39	3.51
Syncarpia glomulifera		0.07	0.58	0.06	0.37	0.22	0.38

Soils are seasonally or generally waterlogged in many areas as a result of irrigation, vegetation removal or other processes. Such waterlogging is assessed or determined in a variety of ways, such as depth to a watertable, soil water content, or presence of other evidence such as mottling in the soil profile. Salinity may be within the soil profile or related to the saturated, possibly fluctuating, zone.

The degree of plant damage and growth reduction will vary with duration of the waterlogging or flooding event, with species and with other interacting factors (for example degree of salinity or temperature). Waterlogging reduces shoot growth by slowing production of new leaves, reducing growth of existing leaves and/or by causing premature leaf shed. It also reduces root growth and viability, and reduces root formation. The above effects are mainly related to the direct effects of low soil oxygen concentrations and changes in plant hormone levels. Woody (older) roots are much more tolerant of flooding than non-woody (younger) roots. High watertables and sodic soils (especially those with a restricting layer) tend to restrict root growth to the surface horizons and this predisposes trees to toppling and, in dry seasons, to drought injury. Prolonged waterlogging can predispose roots to damage from soil pathogens such as phytophthora (*Phytophthora cinnamomi*) and other biological problems. Further, nutrient uptake (for example of nitrogen) is reduced when waterlogging occurs, both as a result of changes in tree physiological processes and changes in soil such as lower mineralisation.

Waterlogging has been reported to produce a range of effects on trees. In *E. marginata*, waterlogging resulted in accumulation of tyloses in tap roots and transpiration rates were related to the level of occluded vessels in the tap root (Davison and Tay 1985). Both *E. camaldulensis*

Table 3.7 Survival (at 12 weeks), relative growth (relative to non-saline control in each case at nine weeks), and sodium, chloride and potassium concentrations in five species of *Casuarina* (van der Moezel *et al.* 1989a).

Species	Treatment	Survival (%)	Rel Ht	Na (ppm)	Cl (ppm)	K (ppm)
C. cristata	Non-saline drained	-	100	3 700	16 300	19 800
C. cristata	Saline drained	100	53	8 600	21 500	10 200
C. cunninghamiana	Non-saline drained	-	100	800	11 100	24 900
C. cunninghamiana	Saline drained	71	36	10 600	38 200	12 900
C. equisetifolia var. equisetifolia	Non-saline drained	-	100	2 600	10 500	17 700
C. equisetifolia var. equisetifolia	Saline drained	100	27	2 000	3 700	11 100
C. equisetifolia var. incana	Non-saline drained	-	100	2 000	8 700	21 400
C. equisetifolia var. incana	Saline drained	100	34	2 200	4 800	12 200
C. glauca	Non-saline drained	-	100	1 800	17 100	22 100
C. glauca	Saline drained	100	34	3 500	9 600	12 300
C. glauca	Non-saline waterlogged	-	84	4 800	14 600	17 600
C. glauca	Saline waterlogged	-	25	13 600	25 600	11 700
C. obesa	Non-saline drained	-	100	1 200	16 500	20 300
C. obesa	Saline drained	99	58	2 800	7 600	9 700
C. obesa	Non-saline waterlogged	-	82	2 600	13 500	16 300
C. obesa	Saline waterlogged	-	46	14 100	26 100	8 700

and *E. globulus* produced relatively more root growth than top growth and more adventitious roots during waterlogging (Sena Gomes and Kozlowski 1980). *E. globulus* was the more sensitive. Clemens *et al.* (1978) showed the order of tolerance to flooding was *E. grandis* > *E. robusta* > *E. saligna* and this was related to the capacity of the species to produce adventitious roots. Plants which tolerate waterlogging usually have specially adapted roots, for example *E. camaldulensis* has root air channels (aerenchyma). The ability of species to withstand waterlogging was related to the development of aerenchyma tissue in *E. viminalis*, *E. ovata* and *E. robusta* (Ladiges and Kelso 1977; Clemens and Pearson 1977).

If saline soils are also waterlogged, non-halophytes will be less able to restrict salt uptake due to lower soil oxygen supply. Salt exclusion mechanisms rely on high metabolic activity and

Table 3.8 Relative growth (%) where control was 100% (Bell and van der Moezel 1988).

Species	Waterlogging	Salinity	Waterlogging/salinity
E. camaldulensis	90	28	29
E. comitae-vallis	12	54	14
E. kondinensis	23	45	10
E. lesouefii	22	51	13
E. platypus	17	55	15
E. spathulata	16	75	15

thus on good soil aeration. Plants under waterlogged conditions must tolerate the stress due to reduced oxygen in the rooting zone in addition to presence of salt. Such a situation means that although there is apparently excess water, there may be problems of taking such water into the plant due to osmotic potentials. Tomar and Gupta (1985) specifically separated saline and aeration conditions. Many highly tolerant species in fact handle both aeration and salt.

The interaction between such factors has not been studied in detail in relation to forest tree species. Van der Moezel *et al.* (1989a) studied five *Casuarina* species in glasshouse conditions in relation to salinity and waterlogging over 12 weeks. Salinity tolerance of *C. obesa*, *C. glauca* and *C. equisetifolia* was associated with exclusion of sodium and chloride while relatively sensitive species, *C. cunninghamiana* and *C. cristata*, accumulated salt in the shoots. Waterlogging with saline water increased salt uptake to the shoot. *C. glauca* was the most tolerant of saline waterlogged conditions (Table 3.7).

Poor substrate aeration is known to decrease salt tolerance in many plants (Barrett-Lennard 1986), with the combination of salinity and waterlogging often being associated with a decreased ability of roots to exclude ions. Marcar (1993) indicated even short-term waterlogging (5–10 days) on moderately saline field sites severely reduced survival of salt tolerant eucalypts. This therefore indicates the need to manage irrigation carefully in saline environments.

In glasshouse studies of stresses due to waterlogging and salinity, there was good survival for waterlogging alone and salinity alone, but survival was significantly affected when treatments were in combination, and relative growth rates were variable (Table 3.8). Survival of *E. camaldulensis* and *E. platypus* was more than 75%.

Stomatal conductance, net photosynthesis and transpiration were studied in 3-month-old *E. camaldulensis* and *E. lesouefii* (van der Moezel *et al.* 1989b). Under non-saline conditions, waterlogging induced stomatal closure in both species. However, the stomates of *E. camaldulensis* re-opened after five weeks when adventitious roots were produced. Relative to that of controls, height growth of waterlogged seedlings was greater in *E. camaldulensis* than *E. lesouefii,* as were rates of photosynthesis and transpiration. In a freely drained medium, high salinity reduced rates of seedling height growth and photosynthesis. In both species, height growth, stomatal conductance and photosynthetic rates were lowered when in saline waterlogged conditions.

The interaction between salinity and waterlogging was assessed in seedlings of four tree species. Where there was no salinity or waterlogging, the growth results were set at 100% (Table 3.9). There was a significant interaction between salinity and waterlogging, and species and their growth.

Table 3.9 Relative height growth (%) of seedlings of four species of eucalypts in saline and waterlogged conditions (Marcar 1993).

		Nil salinity	Salinity
No waterlogging	*E. camaldulensis*	100	83
	E. tereticornis	100	100
	E. robusta	100	70
	E. globulus	100	109
Waterlogging	*E. camaldulensis*	83	72
	E. tereticornis	76	64
	E. robusta	75	54
	E. globulus	72	41

Farrell *et al.* (1996a) assessed the ability of six clones from five provenances of *E. camaldulensis* to produce biomass and utilise water. One clone from Western Australia (M80) produced the greatest total biomass under conditions of waterlogging and increasing salinity, while another clone from the same area had among the lowest production, illustrating the very high variability within the provenances. The response of clones to alkalinity was comparable to that under waterlogging–salinity stress. Water use efficiency was closely related to biomass production. Akilan *et al.* (1997a, 1997b) indicated that clones of *E. camaldulensis* can be screened to be site specific in relation to waterlogging and/or salinity. Tolerance to waterlogging and salinity occur according to different mechanisms, hence selection needs to consider two problems both individually and in combination. Karschon and Zohar (1975) compared partially waterlogged and non waterlogged with increasing salinity and for provenances of *E. camaldulensis* (see Fig. 3.4). Partial waterlogging at low salt concentrations produced higher growth, but at higher salinities there was a major decrease in growth mainly due to leaf loss.

It may be expected that establishment of forest trees at reasonable densities may utilise sufficient water to reduce the level of waterlogging in soils. This may be accelerated by appropriate management techniques such as mounding. The net result will be changes over time in the level of waterlogging and associated salinity.

Concluding comments

The processes by which forest trees tolerate high levels of salt are poorly understood. It is reasonable to conclude that genotypes vary in their ability to avoid and exclude sodium and chloride, and if taken up, where these elements are to be found. The levels which each genotype can tolerate vary within their tissues and with the particular environment. Soil waterlogging increases uptake of salts and the ability of trees to tolerate them.

Systems to assess tolerance in young and old plants are inconsistent, and while reconfirming that well known species such as *E. camaldulensis* are reasonably salt tolerant, standard selection systems are not available to readily assess new genotypes.

The basis for genetic improvement for salt tolerance is selection of several parameters, but a reliable and broadly accepted assessment procedure is required. Chemical analysis of foliage is important but requires careful calibration.

CHAPTER

4

SPECIES SELECTION AND PRODUCTIVITY

Issues

The development of commercial plantations requires the selection of the best genotypes to meet objectives, in addition to appropriate site evaluation and management. Typical selection requires testing of species and provenances and selection from these for improvement. Improvement traditionally takes a long time, but the time can be reduced using biotechnological techniques. Such techniques are critical for selection of genotype and establishment of saline areas.

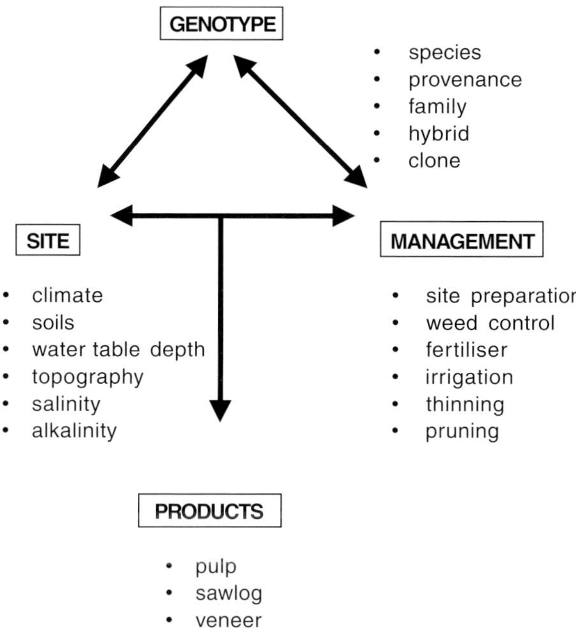

Figure 4.1 Schematic representation of the interaction between site, genetics and management in the development of products from commercial forest plantations.

Forest productivity

The level of productivity, specifically productivity per unit area, is one of the most significant factors in the successful development of commercial forestry plantations. Productivity is affected by three components, namely, site characteristics, genetic composition and type, and level of management inputs. These three elements need to be optimised to produce a specific and defined product. Management inputs include soil preparation, weed control, plant spacing, fertiliser applications, and irrigation on some sites. The factors are interrelated, so when planted on an inappropriate site and/or managed incorrectly the potential productivity of selected genetic material will not be achieved. Species must be selected with due regard to site while the necessary optimum management is determined by both species and site. Such an interrelationship may be considered as a triangle with the planned product being the output (Fig. 4.1). The more intensive the planned management and the greater the effect of limiting factors on growth (such as elevated salinity and waterlogging), the greater the requirement for understanding of interactions between the factors through integrated research.

It needs to be further recognised that the factors of site, genetics and management may change with time, both within a rotation and between rotations, and that the rate of change varies with the factor (Fig. 4.2). Some factors, for example topography, remain static and others, such as rainfall, vary in a random or non-predictable manner and this adds variability to long-term productivity. Some of the changes are site dependent but may change significantly within the period of a rotation. Such factors include soil organic matter, depth to watertable, and distribution of nutrients in the soil profile. The result is that site characteristics change and optimum management for one rotation may not be the optimum for the subsequent rotation.

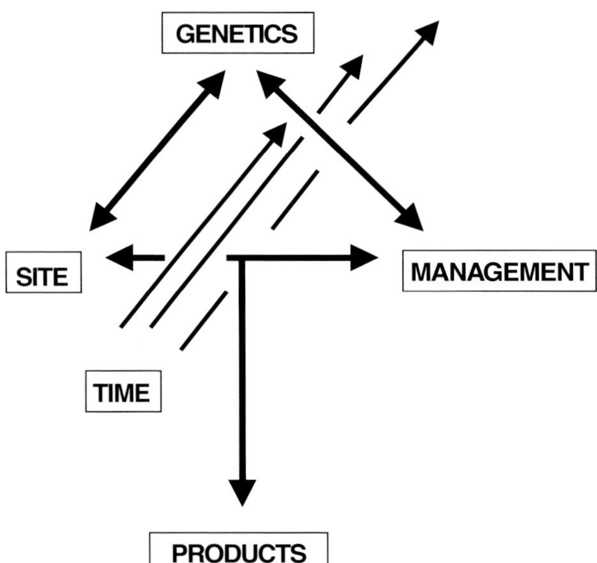

Figure 4.2 Interactions between factors for plantation management as in Figure 4.1 showing the relationship between changes and time.

Genetic stock will be improved over time and will change between rotations. All such factors affect long-term productivity and sustainability, and hence long-term products. An aim of establishment of plantations on saline and waterlogged areas is to result in site changes and management, and hence genetics need to be integrated within such changes.

Measurement of productivity

Forest productivity may be expressed in a number of ways and this is usually related to the proposed end product. The primary measure of forest productivity is the rate of accumulation of organic matter (biomass), that is the quantity of organic matter accumulated per unit area per unit time. This rate of accumulation changes over time and hence the period over which productivity is assessed needs to be indicated (Box 4.1). Such a measure estimates the net productivity of above-ground organic matter without considering replacement of excised tissue (litterfall) and/or below-ground productivity in roots. For commercial purposes, productivity needs to be subdivided since only a proportion of total productivity is utilisable. Hence, in commercial plantations, productivity is usually cited in terms of production of the merchantable product and is normally expressed in t/ha/yr or m^3/ha/yr. It varies depending on location, species, age and product. Estimates of total biomass are of additional importance as they are basic information for estimation of environmental impacts such as nutrient accumulation and removal.

There are a number of species that can be considered for plantation establishment in saline areas but the primary focus here is on the development of *Eucalyptus* and *Acacia* plantations. Most of the available data are for eucalypts planted as exotics in a number of countries. Selection of eucalypts in many areas has been on the basis of potentially high or relatively high rates of productivity. While high productivity figures often quoted for specific

Box 4.1 Estimation and reporting of forest plantation productivity.

Plantation productivity

Net productivity of plantations can be estimated by measuring the accumulation of organic matter and may be considered as the quantity accumulated in a specified period of time. The rate of accumulation changes with increasing stand age and hence both the plantation age and accumulation time periods need to be defined. A typical pattern of accumulation is shown in the following figure (A) based on *Eucalyptus grandis*.

When considered on an annual basis, two estimates of interest are current (periodic) annual increment (pai) and mean annual increment (mai). The mean annual increment is the average rate of accumulation estimated as the total amount of accumulated organic matter divided by the age. The current annual increment is the amount of biomass accumulated in any year (for example, between ages n and $n + 1$ years) and the periodic annual increment is the annualised estimated productivity between two time periods. That is, mean annual increment = biomass at age x divided by x and current annual increment = biomass at age y minus biomass at age $y - 1$. The periodic calculation is over a longer time period than 1 year by dividing by the number of years involved. Mean annual increment and current annual increment are functionally related as shown in the following diagram (B), and the point where they cross is the point where mean annual increment is at a maximum.

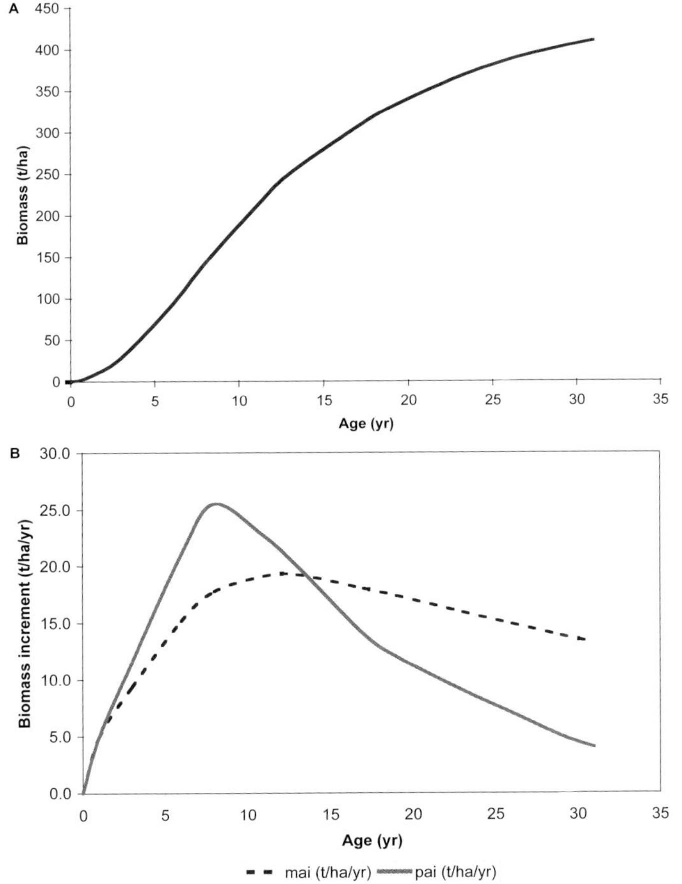

Productivity is often reported in terms of the mean annual increment and it is critically important that the period of time that it relates to is stated. From the figure above, the mean annual increment over 10 years is approximately 19 t/ha/yr, while it is 13 t/ha/yr in the same stand when measured over 30 years.

As biomass is often difficult to estimate, easier measures such as stand basal area or volume are used as surrogates, and curves can be developed representing total or mean annual accumulation over time. As productivity changes with time, the method of expression needs to be specifically stated. In the following diagram, the total volume accumulation is 73 m^3/ha for *E. camaldulensis* Site Quality I at age 4 years, the mean annual increment is 18 m^3/ha/yr, and the current annual increment is 33 m^3/ha/yr. All of these are functionally related.

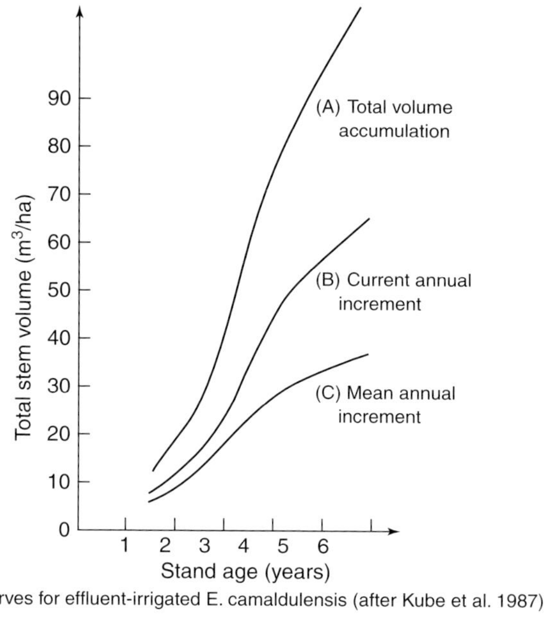

Yield curves for effluent-irrigated E. camaldulensis (after Kube et al. 1987)

locations are based on individual plots, the average productivity of extensive plantations is often much lower and more variable. The reasons for this fall down from research or demonstration trials are not always immediately obvious but generally are due to inadequate site assessment, poor management procedures and insufficient evaluation of genotypes. When estimating and comparing productivity from different areas, the rotation length is a key element. Because of the nature of growth curves, mean annual productivity will appear higher in short rotations, so data for different ages are not truly comparable. Overall levels of productivity and rotation length have been estimated as an indication (Table 4.1).

The minimum acceptable average productivity for a project will be determined by factors such as the costs of inputs and the value of products, and hence there will be considerable variation of acceptability. However working estimates of productivity for rotations less than 15 years of age are a minimum of 20 to 25 m^3/ha/yr mean annual increment and, probably, longer rotations have a minimum requirement of 15 m^3/ha/yr. Lower figures may be accepted when other values such as environmental improvements, which may relate to soil or water, are included.

Table 4.1 Profile of *Eucalyptus* spp. plantation growth in different countries (Schreuder *et al.* 1990). The rotation lengths represent the proposed end products from the plantation.

Country	Rotation age (yr)	Productivity mean annual increment (m^3/ha/yr)
Brazil	6–7	25–55
Portugal	8–10	4–40
Argentina	6–18	18–30
Chile	6–18	18–30
Spain	10–12	4–49
South Africa	15–20	18
Australia	30–45	12

Table 4.2 Selected production regimes for *Eucalyptus* on the NSW north coast (Knott 1996). The timber values are present day values.

Stand age (yr)	Product	Wood yield (m^3/ha)	Nominal value ($A/$m^3$)
12	Pulpwood	34	8
	Small salvage sawlogs	34	13
20	Pulpwood	25	9
	Small sawlogs	88	25
30	Pulpwood	63	11
	Small sawlogs	135	31
	Large sawlogs	69	56
	Veneer logs	116	69

In any management regime, other than entirely for pulp production, there will be a mixture of products which can be manipulated to optimise value. One example of a regime is shown in Table 4.2 for *Eucalyptus* plantations in northern NSW grown on a 30-year rotation. Most products are obtained at the clearfall stage where the unit value of the product is much higher, but on this schedule there is an extended period for return on investment, compared with harvesting for pulp at a much earlier age.

In addition to the total productivity for a given area, the size of individual trees at a given age is critical in terms of their merchantability. To obtain a value for such a stand, the determination of tree size distribution at different ages needs to be analysed. Tree diameter for a given site and age is affected by the number of trees established per unit area. High numbers (1500+ stems/ha) used for pulp will give high productivity per initial unit area but smaller individual tree size, whereas low numbers (for example < 800 stems/ha) will give low productivity per initial unit area but larger individual tree size. There are further interactions with the genetic material selected. Three seed sources of *E. grandis* grown in Queensland (Table 4.3) showed variation in size class distribution. The Atherton seed source, considered to have significant in-breeding, produced a large number of small trees while there were relatively fewer small trees in the Pomona provenance and the Coffs Harbour seed source.

Table 4.3 Size class distribution for three seed sources of *E. grandis* grown in Queensland. The seedlots were considered to have varying degrees of inbreeding with Atherton being unimproved/wild, Pamona plantation unimproved with outcrossing, and Coffs Harbour unimproved seed orchard (Ryan 1993b)

	Seed source		
Diameter class (cm)	Atherton	Pomona (% of trees)	Coffs Harbour
5.0–8.9	4	0	0
9.0–12.9	21	13	7
13.0–16.9	18	19	23
17.0–20.9	32	41	23
21.0–24.9	21	28	47
25.0–28.9	4	0	0
Total	100	100	100
Average	15.1	16.4	17.4
Minimum	6.3	7.8	9.9
Maximum	29.5	27.7	22.5

In terms of selection, there would be more trees available to develop a commercial solid timber crop from the Coffs Harbour seed source taken from an outcrossed seed orchard.

The selection of appropriate species determines the quality and quantity of the final product. The selection of appropriate plants (genotypes) for a site involves determination of suitable species, provenances, hybrids, families or clones; and a genotype is also a determinant in selection of sites and management procedures. The suitability means that the products fulfil commercial requirements and that they are also able to grow on the site. Performance of species is usually assessed from trial field plantings and the initial parameters are survival and growth followed later by volume or biomass assessments. Where a commercial crop is to be grown, the growth rate of the merchantable component of the stand will be of primary interest. Hence, both total biological production and actual distribution of the organic matter are critical issues. Ideally testing of a species would be carried out in relation to a gradation of environmental conditions to ascertain resilience of species under change. Such assessments may be undertaken in relation to factors such as topography (see Box 4.2), rainfall, slope, soils, salinity and/or waterlogging. However little information is available in relation to environmental gradients as most studies are undertaken as comparisons between species and on only one site.

The changes in site conditions and properties over time need to be considered. Such changes are related to soil chemical and physical attributes, and hydrological conditions. These changes will affect the productivity of plantations over time. For example, decreases in soil salinity over time within the profile will affect the interpretation of tree tolerance to salinity. The assessment of relatively mature trees in a saline environment will need to consider the conditions at the time of tree establishment. Further, salt may decrease or increase with time, and with increased size trees are extending roots into more or less saline areas. Such temporal changes need to be carefully measured and documented in any trials that are established.

Box 4.2 Use of performance trials for evaluation of species across environmental gradients.

Species and provenance trials allow the performance of the species to be assessed, and with repetition over a series of environmental conditions, consistency of performance can be assessed. Species may perform well across a range of sites or may be very site specific. Performance in a trial on the north coast of NSW demonstrates the effect of narrow and broad changes in environmental conditions. There was a narrow change in topography between the two sites at Cascade and a change in general environmental conditions at Yabbra.

Site	Cascade 1	Cascade 2	Yabbra
Elevation (m)	640	640	520
Rainfall (mm)	1560	1560	1080
Geology	slates	slates	sandstone
Topography	moist ridge	moist gully	dry ridge
Basal area growth at 14 years of age (m^2/ha)			
E. agglomerata	30.6	20.8	11.2
E. cypellocarpa	5.2	9.0	27.8
E. dunnii (Provenance 1)	31.2	36.8	33.9
E. dunnii (Provenance 2)	34.0	33.8	not planted
E. grandis	**35.2**	**24.1**	**40.4**
E. pilularis	*36.8*	*failed*	*28.7*
E. saligna (Cascade seedlot)	22.8	23.1	18.9
E. saligna (New Zealand seed)	5.8	8.5	9.1

Species such as *E. dunnii* and *E. grandis* have performed reasonably well across the three sites tested. *E. dunnii* showed minor differences between provenances. The performance of *E. saligna* has been poor to moderate across all sites and there is a large difference between seedlots indicating the value of evaluation of individual provenances or seedlots. Seedlots performing well elsewhere (for example improved *E. saligna* from New Zealand) may not perform well under other conditions. Species such as *E. pilularis* are much more site specific producing the best growth on one site while totally failing on another. These forms of trials indicate the need for extensive field testing of genetic material.

From Johnson and Stanton (1993).

Development of forest crops on saturated and salinised soils will require the use of tolerant genotypes. Survival and acceptable growth are the main assessment criteria, but for commercial establishment analysis of total productivity is required. Rankings of species for salinity tolerance have been developed for a number of Australian genera and species including *Acacia*, *Casuarina*, *Eucalyptus* and *Melaleuca* (Blake 1981; El-lakany and Luard 1982; van der Moezel and Bell 1987; van der Moezel *et al.* 1988; Marcar 1989; van der Moezel *et al.* 1989a; Craig *et al.* 1991; van der Moezel *et al.* 1991). Many studies recognise that the combination of salt and waterlogging is a critical factor but few studies have routinely assessed tolerance of genotypes under this scenario.

Species selection for plantations

Selection of a species for planting depends upon the desired end product and the potential productivity of alternative species. The selection may be relatively easy if an end product has been defined, for example the specific requirement to use *E. globulus* for pulp. Alternatively, it may require a much greater level of analysis such as where there is a range of potential products. Such a selection process requires that there are a considerable number of studies in the form of field trials. Commercial purposes require that a major decision is made as to whether the end product is to be softwood-based (for example pine) or hardwood-based (for example eucalypts).

Softwoods

The main softwood species planted in the Southern Hemisphere is radiata pine (*Pinus radiata* D. Don) which has been planted in South America, New Zealand, Southern Africa and Australia. Smaller areas of Southern Pine (*Pinus taeda* and *Pinus elliottii*), Caribbean Pine (*Pinus caribaea*) and Maritime Pine (*Pinus pinaster*) are also planted. The plantations have been established to provide a wide range of products including clear sawn products, veneers and reconstituted products. In most cases the original reason for planting was to fill the need for coniferous timbers, these areas not having readily available supplies of material as is the case in many parts of the Northern Hemisphere. Radiata pine and Maritime pine both appear to have moderate tolerance to salinity in terms of survival, but little information is available on the effect of salinity on their rates of growth under field conditions.

Hardwoods

Two of the main genera of hardwoods considered for broadscale fast growing commercial timber production in plantations are *Eucalyptus* and *Acacia*. Eucalypts have several sub-genera but the main focus has been on the species within the sub-genera *Symphyomyrtus* with a few from *Corymbia*, the selection being based primarily on experience with some supporting trials (Pryor and Johnson 1971; Boland *et al.* 1992). Species in the other sub-genera such as *Monocalyptus* and *Eudesmia* have generally performed relatively poorly. Trials have shown that *Symphyomyrtus* species extend across all levels of salinity from low to high tolerance (in relative terms), while *Monocalyptus* species primarily have low salt tolerance (Table 4.4).

Field testing of species needs to consider environmental variability including spatial variability such as soils and topography, or temporal effects such as annual rainfall variation. Such testing will indicate whether species perform well over a wide range of conditions or only under very specific conditions. An outcome will be the ability to select species for specifically identified environmental conditions. Testing of species needs to be across critical gradients. For example, trees were tested in one study in saline and adjacent environments over a range of rainfall regimes (800, 520 and 420 mm per annum). When assessed at 7 years of age, the study provided information on growth in relation to salinity and drought, and based on leaf area and crown volume, the best performers out of 25 species were identified (Table 4.5). With regard to survival and salt tolerance, growth rate was a critical factor. *E. cladocalyx* performed reasonably well over the three sites but most other species were more site specific. For example, only *E. globulus* and *E.(C.) maculata* were the better performers at the site with 800 mm rainfall.

A large number of trials have been carried out on species performance under a wide range of conditions throughout the world. In areas being developed for forest plantations, it is a usual procedure to test the performance of potential species and provenances under the particular

Table 4.4 Performance of species of *Eucalyptus* in relation to salinity, ordered according to sub-genera (Pryor and Johnson 1971; Blake 1981). The salinity level is set broadly where 50% of the seedlings have been killed.

	Salt conc. (mM) LD_{50}	*Corymbia*	*Eudesmia*	*Monocalyptus*	*Symphyomyrtus*
High tolerance	450+				*E. woodwardii*
	400+	*E. calophylla**	*E. erythryocorys*		*E. tereticornis*
	400+				*E. camaldulensis*
	380+				*E. largiflorens*
	370+				*E. grossa*
	350+				*E. neglecta*
	320+				*E. incrassata*
	320+				*E. globulus* spp. *globulus*
	320+				*E. lehmanii*
Moderate tolerance	300+	*E. ficifolia**			*E. globulus* spp. *bicostata*
	300+				*E. blakleyi, E. torquata*
	280+				*E. tetraptera*
	250+		*E. tetragona*	*E. alpina*	*E. astringens*
	250+				*E. gomphocephala*
	230+				*E. caesia*
	225+				*E. campaspae*
	220+				*E. goniocalyx*
	220+				*E. spathulata*
	210+				*E. leucoxylon*
	210+				*E. sideroxylon*
	210+				*E. bosistoana*
	210+				*E. cladocalyx*
Low tolerance	200+			*E. obliqua*	*E. behriana*
	200+				*E. microcarpa*
	190+			*E. globoidea*	*E. lesouefii*
	190+			*E. elata*	
	190+			*E. macrorhyncha*	
	185+				*E. albens, E. botryoides*
	185+				*E. kondininensis*
	180+	*E. maculata**		*E. muellerana*	
	170+			*E. sieberi*	
	170+			*E. radiata*	
	160+				*E. aggregata*
	160+				*E. crenulata*
	150+			*E. stellulata*	
	150+				*E. occidentalis*
	140+			*E. baxteri*	
	130+			*E. regnans*	*E. salubris*
	130+				*E. stricklandii*
	130+				*E. oxymitra*

* now *Corymbia calophylla, C. ficifolia, C. maculata*

environmental conditions. By establishing trials over a range of sites, assessments can be made of growth in terms of individual sites and across a range of conditions. Typically it is found that some species are very site specific, while others tolerate a much broader spectrum of conditions (see Box 4.2). This raises the issue of alternate strategies, that is selection of one species to perform well over a broad set of conditions or selection of a different species for each specific condition.

Table 4.5 Best species performers out of 25 species at 7 years of age over different rainfall regimes (Biddiscombe *et al.* 1981, 1985b).

Annual rainfall (mm)		
880	**520**	**440**
E. globulus	*E. occidentalis*	*E. leucoxylon*
*E. maculata**	*E. cladocalyx* x var. *mana*	*E. cladocalyx*
E. cladocalyx	*E. sargentii*	*E. astingens*

* now *Corymbia maculata*

Many of the large number of trials on trees grown and tested under saline conditions overseas have been principally for the objective of land reclamation. Evaluation has usually been for land care catchment protection or social programs, rather than commercial plantations (Morris 1984). Survival is the primary index of success rather than growth, and it is often difficult to generalise from the data since in many studies insufficient environmental information is provided. However, the information provides 'key species' information for future research development especially in extreme situations. Examples include satisfactory growth of *Eucalyptus* 'hybrids' in India where the pH of surface soil is less than 9 and soluble salts are less than 0.3%. However *Eucalyptus* 'hybrid' failed where pH was greater than 10 and surface soil soluble salts by weight were greater than 0.7% (Yadav and Singh 1970). This study indicated extreme conditions for growth of *Eucalyptus*. In Israel, it was reported that *E. camaldulensis* was tolerant of pH 7.0–8.2 and tolerant of 'high' salinity where the site was high in calcium carbonate with chlorides and sulfates present (Karschon 1966). The Israeli study raised the issue that depending on the form of salt, there may be differing results. Jain *et al.* (1985) in Rajasthan tested a range of species in field trials with varying levels of salinity and varying seasonally. *E. camaldulensis* produced the greatest growth. Sardabi (1991) reported on trials on one alkaline and one poorly drained site in Iran. At the alkaline site the best species were *E. dalrympleana, E. viminalis, E. camaldulensis, P. eldarica* and *P. pinea* while on the poorly drained site, *E. dalrympleana* and *E. camaldulensis* performed best.

The provision of such information provides very broad indicators only, possibly directing attention towards potentially more tolerant species. The general indication is that *E. camaldulensis* is tolerant of a wide range of conditions but there is a high level of variability within the species. The value is that it indicates the form of salt causing the higher conductivity is a critical factor and needs to be considered in evaluation trials. Marcar *et al.* (1991a) provided three-year results from plantings of potentially commercial species on saline discharge areas. The recommended species included *E. globulus*, *Acacia melanoxylon* (low to moderate salinity sites (EC_e < 10 dS/m), *E. camaldulensis* and *Casuarina glauca* (moderate salinity EC_e 10–20 dS/m), and *Melaleuca halmaturorum* (high salinity EC_e > 20 dS/m). Treatments such as protection, mulching and fertiliser have increased survival and growth. Similarly, Marcar *et al.* (1991b) carried out similar analyses in the more tropical areas of Pakistan and Thailand. Again, species tolerant of moderate salinity included *E. camaldulensis*, *E. occidentalis* and *Acacia saligna* (EC_e 10–20 dS/m) while species tolerant of moderate to high salinity were *A. ampliceps, A. stenophylla, A. maconochieana, Casuraina glauca* and *Melaleuca leucadendra* (EC_e > 20 dS/m).

Generally studies in a number of countries show that *E. camaldulensis*, selected as a timber production species, survives well over a broad range of saline environments but that other species, such as *E. globulus*, perform well in specific generally low saline environments.

Table 4.6 Published varieties of *E. camaldulensis* as summarised by Bennett and George (1996).

Variety	Published location
E. camaldulensis var. *acuminata* (Hook.) Blakely	–
E. camaldulensis var. *brevirostris* (Mig.) Blakely	–
E. camaldulensis Dehn. var. *camaldulensis*	Lake Albacutya, Vic.
E. camaldulensis var. *obtusa* Blakely	Gilgunnia, NSW, Kuitpo, SA, Alice Springs, NT, Strelly River, WA
E. camaldulesis var. *pendula* Blakely and Jacobs	Daly Waters, NT
E. camaldulensis var. *subcinerea* Blakely	Silverton (near Broken Hill), NSW Charleville, Qld

Genetic selection and breeding programs

The primary objective of breeding programs is to improve the desirable characteristics within a species in order to increase the value or utility of the crop. Such characteristics include higher productivity, increased value through stem straightness, reduced branch size, modified wood properties, increased insect resistance, and/or modified oil contents.

Breeding and domestication programs for eucalypts are described in detail by Eldridge *et al.* (1993), providing examples of improvement from various forestry programs. Traditional selection programs are initially hierarchal ranging from broad analyses of species and provenances through to narrower and more explicit analyses of families, hybrids or clones. At the first level, selection of the best performing species is determined through species selection or elimination trials (see Box 4.2). Within a species there is considerable variation in plants produced from seedlots obtained from different locations and as such, they provide sources of basic genetic material which can be used for improvement. *E. camaldulensis* has the widest natural distribution of the eucalypts providing substantial genetic variability which has been utilised in breeding programs (Midgley *et al.* 1989). In 32 well-designed field trials in Mediterranean and tropical countries, growth and properties of *E. camaldulensis* seedlots were tested and results from 24 trials of 8- to 10-year-old stands were reported by Lacaze (1978). Within the winter rainfall zone (Mediterranean climate), the Lake Albacutya Victorian provenance was outstanding in every trial while in the tropical summer rainfall zone, the Petford (Queensland) and Katherine (Northern Territory) were superior in growth and straightness. This indicated the genotype x site interaction occurred at the broadest level but not within a broad environmental zone. Bennett and George (1996) reported on the performance of *E. camaldulensis* in trials in Western Australia in which they recognised six botanically distinct varieties which gave different performances (Table 4.6). Varieties of *E. camaldulensis* have been the most salt tolerant in many trials but this is not universally so, and is partly a reflection of differences between seedlots.

After initial analyses there is further improvement through breeding programs, which involve selection of superior individuals from base populations and development of a breeding population. The individuals are vegetatively reproduced for introduction into seed orchards where there may be either general crossing or controlled pollination. The seeds from such crosses are then tested in progeny trials and the best crosses further selected. For overall and continued improvement a selection program is maintained, and at any stage

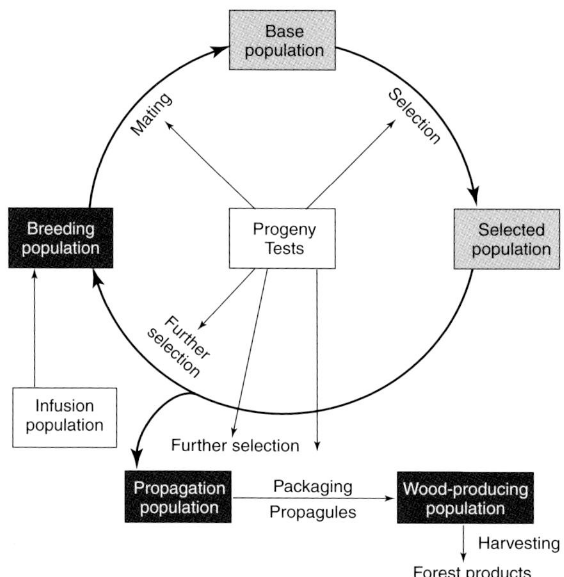

Figure 4.3 Generalised cycles outlining a genetic improvement program and delivery system for commercial forest plantations.

within this superior individuals are used in the general plantation program. Propagation for operational planting can be through production of seed or through a variety of vegetative means. The above processes go through a series of cycles as summarised in Figure 4.3 but require long-term planning and management to ensure continued success.

A number of strategies are involved in such breeding programs reflecting the type of material available for the breeding program and the size of the plantation program it is supporting. Limitations for such programs include the length of time involved, often decades, before gains are achieved and also the requirement for assessing the genotypes through their phenotypic characteristics. The use of fingerprinting and marker aided selection can accelerate and hence improve the efficiency and gains of such programs as they provide genetic information (Fig. 4.4). This is achieved by more precise identification of superior parents.

Provenances

The term provenance is used rather loosely to define seedlots of a species obtained from different localities or environments in natural stands. Eldridge *et al.* (1993) use the term 'seed source' or 'land race' when seed is collected from planted trees. There is an implicit assumption that there has been adaptation due to differing environmental conditions. The variation within a small area of a natural population may be as high as that over a wider geographical range. Changes arising from slight climatic variations may be relatively minor and require large distances to detect differences, while major soil boundaries may lead to significant differences over small distances. This latter effect may be expected where there are salt effects, that is large differences over small spatial differences. Testing of seedlots needs to be systematic and comprehensive if they are to be used in breeding programs.

Particular species are more tolerant of the severe conditions of waterlogged and saline soils, and there are large variations in tolerance between provenances within species and also within

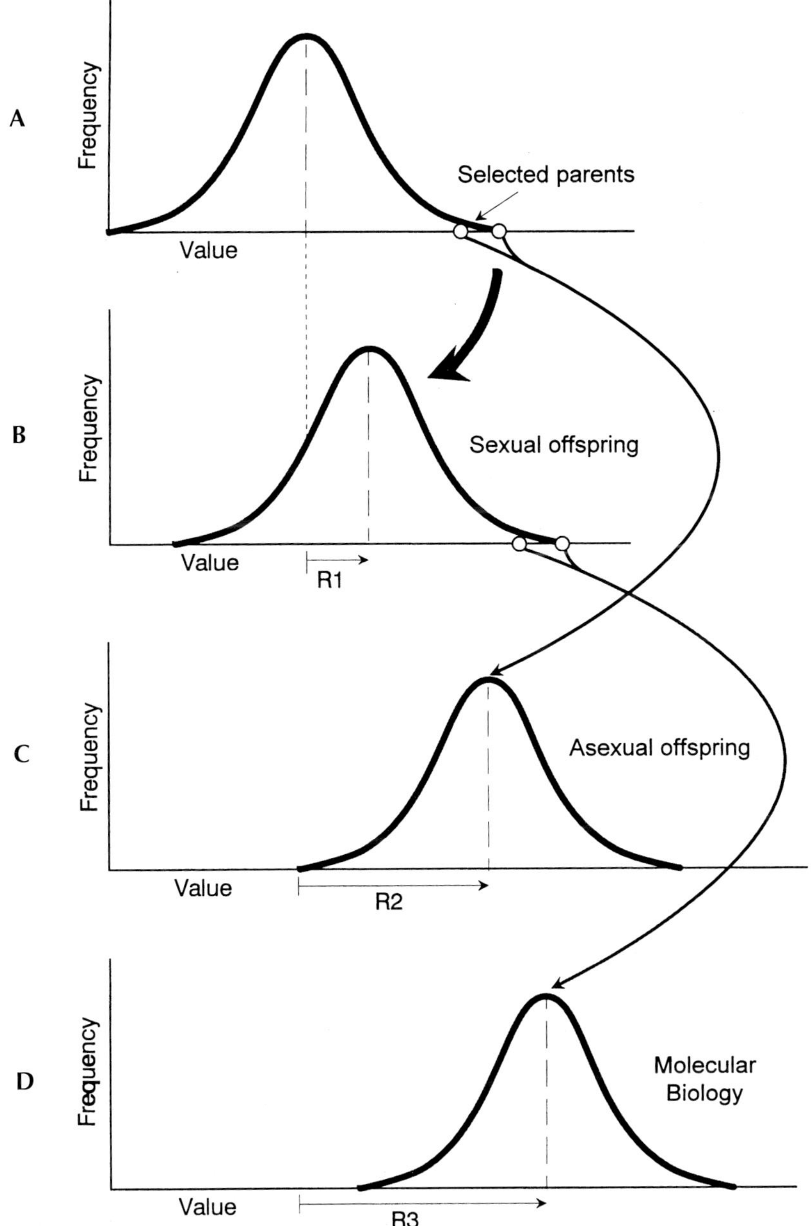

Figure 4.4 Progeny from a number of plus trees may exhibit characteristics (for example chloride tolerance) as shown in A. When those selected in A are mated, the offspring can be expected to have a distribution as shown in B, that is there is an R1 increase in the mean. If the parents had been reproduced by asexual means (clones), an immediate gain (R2) can be expected in the offspring. By use of molecular biology, the expected increases can be even greater (R3). The level of increase at any stage depends on the specific character of interest and the species involved.

Table 4.7 Ranking of provenances of *E. camaldulensis, E. tereticornis* and *E. rudis* in order of decreasing NaCl tolerance (Thomson 1987; Thomson *et al.* 1987).

		NaCl level (M) causing mortality	
Species	**Provenance**	**Mean**	**Homogeneous subsets**[1]
E. camaldulensis	DeGrey River, WA	0.636	a
E. camaldulensis	Silverton, NSW	0.613	ab
E. camaldulensis	Wiluna, WA	0.608	ab
E. tereticornis	Loch Sport, Vic.	0.591	bc
E. camaldulensis	Pentecost River, WA	0.587	bc
E. camaldulensis	Emu Creek, Qld	0.565	cd
E. rudis	Wagin, WA	0.535	de
E. camaldulensis	Quilpie, Qld	0.532	def
E. camaldulensis	Kangaroo Island, SA	0.522	efg
E. tereticornis	Laura, Qld	0.512	efg
E. camaldulensis	Lake Albacutya, Vic.	0.510	efg
E. tereticornis	'Strathfieldsaye', Vic.	0.508	efg
E. camaldulensis	Whiteheads Creek, Vic.	0.507	efg
E. camaldulensis	Minlaton, SA	0.503	efgh
E. rudis	South Yunderup, WA	0.500	efgh
E. camaldulensis	Douglas, Vic.	0.499	efgh
E. tereticornis	Kenilworth, Qld	0.494	fgh
E. camaldulensis	Katherine, NT	0.490	gh
E. camaldulensis	Moree, NSW	0.466	hi
E. tereticornis	Raymond Terrace, NSW	0.435	i

[1] Provenances which do not share a common letter have means which are significantly different at the 5% level.

provenances (van der Moezel and Bell 1990, Marcar 1993). Red gums (*E. camaldulensis, E. tereticornis and E. rudis*) have exhibited high salt tolerance, however there is a very large difference between provenances, so there was no clear distribution at the species level (Table 4.7). This wide range of tolerance by particular genotypes has led to the strategy of selection of individuals, followed by micropropagation of particular individuals.

The performance of genotypes will be affected by the site on which they are planted and the proposed establishment management regimes. In the traditional rainfed plantation areas, a number of studies have emphasised the interaction between these factors. One such study on the productivity of *E. grandis* was set in rainfed areas of sub-tropical Australia. It addressed productivity in relation to establishing *Eucalyptus* plantations in a new area, and hence growth in relation to climate and soils, genetics, and management were included. The factors included bio-climatic analyses of the region, selection of areas climatically suitable for plantations, identification of typical soil types, identification of species, and determination of nutrient requirements for operational levels. It was concluded that the main limiting factors for tree growth were drought and nutrients, and it was further concluded that *E. grandis* was

Table 4.8 Variation in growth between provenances of *E. camaldulensis* treated with saline effluent (Dale 1992).

Provenance/seed source	Hollands Lake Height (m) 21 months	Dows Lane Height (m) 26 months
Lake Hindmarsh, Vic.	2.8	4.6
Silverton, NSW	2.0	
Umberumberka Creek, NSW	1.75	2.6
Alcoa, Clone CML 512	1.2	

the best species over widely differing sites in Queensland and that nutrients were the major limiting factor for that species. Annual productivity reaching 30 t/ha/year in the second year of growth correlated well with mean canopy nitrogen. Different provenances were trialled on three soil types and while none of the soils were saline, the principle of site (soil) genotype interaction and the requirement to select appropriate genetic material for a site in order to achieve optimum results were indicated (Ryan 1993a; Cromer *et al.* 1995).

Eucalypt provenances have been tested under saline conditions using saline effluent treatments. Results for provenances of *E. camaldulensis* indicate the type of variability within species suitable for improvement (Table 4.8). Within this small trial there was at least a two-fold difference between provenances.

Testing of genotypes

Existing breeding programs, particularly improvement of trees in short rotations such as for pulp production, have been very successful especially when integrated with sound site selection and site specific silvicultural management practices. For example, clones and hybrids of *E. grandis* and *E. urophylla* have been established in conjunction with improved management and there have been large increases in productivity and significant advances since that time (Campinhos 1980). The management and environmental conditions under which genotypes are tested significantly affect their performance. Many programs have an objective of producing material which will perform well over a wide range of sites while other programs aim to produce material which will perform well on specific sites (site x genotype interactions). The variability in saline conditions means that site specific genotypes are probably going to be the best option. Various types of field trials need to be designed so that in addition to assessing specific performance, the process involved in saline tolerance or avoidance can also be analysed.

The effects of the interactions between provenances and optimum management treatments have been demonstrated for *E. grandis* grown in Queensland (Cromer *et al.* 1995). Four seedlots were studied in conjunction with factorial treatments of fertiliser and irrigation. At 5.6 years, the stem volumes from untreated trees ranged from approximately 20 to 55 m^3/ha for different provenances (Fig. 4.5), while the differences with fertiliser addition were 140 to 190 m^3/ha, and with fertiliser and irrigation between 120 and 270 m^3/ha.

Without treatment, Coffs Harbour produced the best growth and Atherton the worst. Application of irrigation essentially did not improve growth while fertiliser (nitrogen and phosphorus) greatly increased growth of all seed sources. Coffs Harbour produced the best response to fertiliser application while Pomona gave the best growth with fertiliser and irrigation added simultaneously. The response patterns were different between seedlots. The results

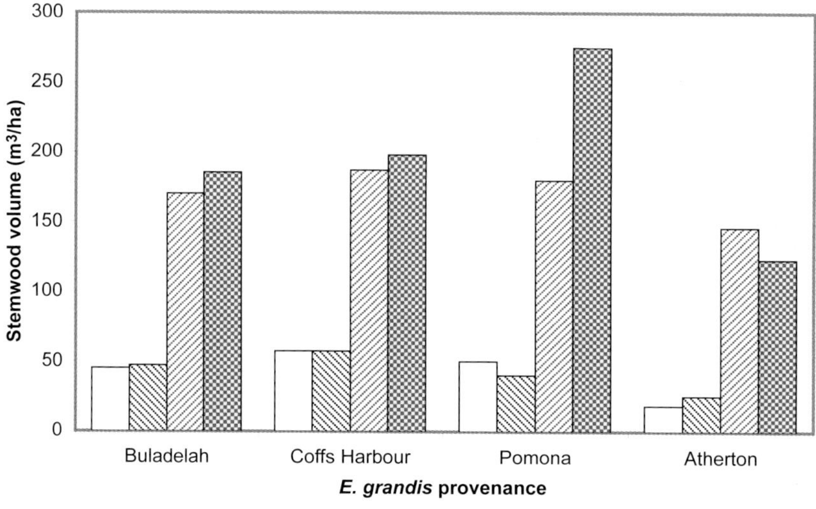

Figure 4.5 Stemwood volume of four seedlots of *E. grandis* subject to fertiliser and irrigation treatments in factorial combination, at 5.6 years after planting (Cromer *et al.* 1995).

Table 4.9 Variation in insect herbivory and related plant characteristics for two populations of *E. camaldulensis* (Stone and Bacon 1994).

	High cineole population	Low cineole population
% Total oil yield (w/w leaf dry weight)	2.12	1.23
% Cineole of total oil yield	71.5	12.74
% Leaf area missing (insect herbivory)	11.4	18.0
Diameter increment (cm)	1.14	0.03
Foliage nitrogen (%)	1.51	1.37

of the fertiliser trial indicated that both phosphorus and nitrogen were limiting to growth. The value of such a trial indicates the need to test genetic material in conjunction with various management treatments, testing not just basic growth but also responsiveness to treatment.

Assessment of productivity and its improvement through genetic selection usually focuses on wood productivity. Other products, such as oils, are also potential products and require separate analysis. Such products are often highly heritable both in content and composition and can be improved through breeding (Bartle 1995). The properties of oils may also be considered in relation to forest production processes, such as insect resistance, in that modified oil levels may change the susceptibility of insect attack. For example, in *E. camaldulensis* two populations were found at one site, one relatively high in 1,8-cineole of terpenoid (oil) yield and the other low. The high yielding oil variety had lower insect herbivory (Table 4.9). There was also high variation in oil content and composition of terpenoids in *E. camaldulensis* and *E. tereticornis*. In *E. camaldulensis*, some provenances were almost devoid of β- and α-pinene (Fig. 4.6) while others had high total and relative quantities. Breeding programs can focus on specific characteristics to improve quantity and quality.

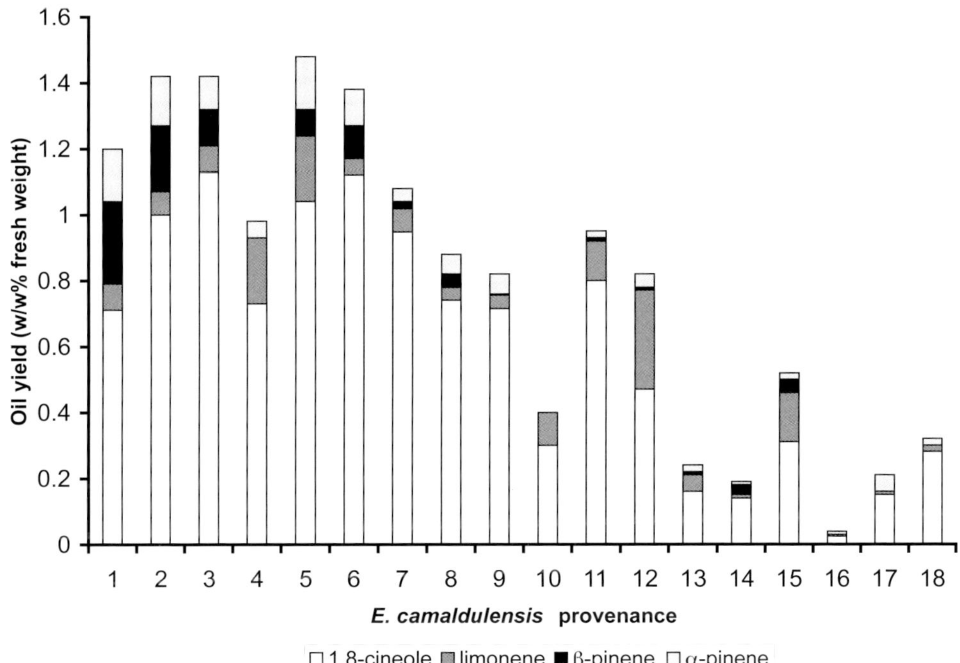

Figure 4.6 Yield of 1,8-cineole, α-pinene, β-pinene and limonene in eighteen *E. camaldulensis* populations as sampled in the wild in northern Australia (Doran and Brophy 1990).

Application of biotechnology to the improvement of trees

The effectiveness of traditional methods of improvement in plantation trees is limited by the long generation times, that is the time to plant, flower, fertilise, produce seed and test the material. These traditional systems of selection and breeding can be improved and accelerated through the use of biotechnology, specifically being a reduction in the generation time for improvement. Such systems can accelerate improvement either within species or via crossing between species to obtain hybrids with specifically identified characteristics. Modern biotechnology techniques offer significant potential to accelerate development of productive, salt tolerant trees. Such techniques include more precise use of natural sources of variation within and between species, introduction of novel sources of genetic variation from unrelated species, and methods of rapidly delivering improvements to the field plantings.

Most breeding assessments consider the observable and measurable characteristics of individuals, the phenotypic characteristics. This provides some insight into genotypes but more precision is obtained by undertaking genetic analysis.

A genetic map of tree species involves the identification of molecular markers throughout the plant's chromosomes. A map is created once all the chromosomes have been 'saturated' with markers. The locations of markers and their relationship to quantitative traits such as salt tolerance or volume growth are termed Quantitative Trait Loci (QTL). The QTL do not define the location of the gene on a chromosome, rather they define the zone of chromosome in which a gene is present. Interpretation and use of the map then depend on whether the presence or absence of markers on a chromosome can be correlated with traits in the tree growing in the plantation (the phenotype). For example, if the presence or absence of markers can be correlated with differences in salt tolerance, then a QTL can be defined. If a QTL for a trait can be established, then the QTL can be used to show activity of the gene in the parents and progeny, which can be selected for further tree improvement or for clonal production.

Using linked markers, different traits can be precisely combined into a hybrid, or introgressed from one species into another. In this way, salt tolerant genes from *E. camaldulensis* can be identified and combined with genes for fast growth and wood quality from *E. globulus* and *E. grandis*. The future performance of progeny produced by marker guided crossing can be predicted while plants are still small seedlings. The time required to realise the results of marker guided breeding can be substantially reduced relative to conventional breeding by identification of superior seedlings based on the genes they carry, and vegetative propagation of these seedlings.

Genetic marker guided breeding makes it possible to compress decades of conventional breeding into a matter of years, and some applications, such as the introgression of salt tolerance genes from one species into another, may be compressed from around 30 years to three or four years.

Genetic mapping

Genetic mapping provides a means to more precisely and rapidly exploit natural sources of genetic variation than can be achieved by conventional breeding alone. This technology enables identification of markers linked to genes controlling traits of commercial interest such as growth rate and timber quality, as well as traits for survival under stressed conditions such as salinity and disease. This in turn enables the future performance of seedlings to be predicted based on the genes they carry, rather than their external appearance (phenotype), and hence there is early deployment of superior genetic material. Where no useful natural variation for a desired trait exists within a species of interest or in a closely related species, genetic engineering offers the ability to find useful genes in unrelated species and to move these genes across species barriers.

This approach can be extended to wide crosses, where one species may be of no commercial value itself, but may be able to contribute a valuable trait not found in the productive timber species, such as high salt tolerance or disease resistance. For example, genetic mapping may be able to be used to identify salt tolerant genes in hybrids between *E occidentalis* (which has no commercial value as a timber species) and *E. globulus* (a highly desirable timber species). Using marker assisted back crossing of the hybrid to *E. globulus* progeny, then predominantly *E. globulus*, but carrying only the genes of interest from *E. occidentalis*, may be selected. This process can be achieved in a few years using biotechnology methods, but would take at least 30 years by conventional breeding approaches. Eucalypts with mallee forms and which have extreme resistance to salt are not feasible for inclusion in traditional tree improvement programs but may be included when using Marker Aided Selection and genetic engineering. For example, species such as *E. occidentalis*, *E. sargentii* or *E. diptera* are species which are not normally considered for plantations but may provide key genetic material. Such technologies are limited where species have low commercial value or there is only a small market for the product because of the relatively high cost of field and laboratory testing.

Marker aided selection

Conventional tree breeding identifies superior genotypes (parent clones) by observing their performance (plus trees) or the performance of their offspring. If a specific genotype or clone is to be tested it must be propagated (either vegetatively or through seed) and established in field trials. The evaluation requires a number of years (minimum of five for growth and probably more for wood properties). Alternatively, marker aided selection (MAS) can be used to identify trees with superior performance through examination of bands produced by their DNA: that is, trees with specific bands may perform better than trees with a different banding pattern. The actual markers used for selection are not the genes controlling performance themselves, but the two are close enough on the chromosome (that is they are genetically linked) that the marker and the control genes are passed on together. The molecular markers can be measured in very young material and hence, if the correct markers are known, the time for selection and evaluation is greatly reduced. While the system is expensive and requires specialised knowledge, the primary benefits are in precision of selection and the gain in turnover time. Where markers for increased aspects of salt tolerance can be identified, the selection procedure for this trial can be improved.

Genetic engineering

Genetic engineering or genetic modification of eucalypts has been undertaken by a number of organisations. A modified plant has genes that are not present in the original gene pool, and traits that are difficult to breed for. For such genetic modification, critically the trait of interest is regulated by the activity of only a small number of characterised genes. The products of such development need to be carefully managed and would generally be reproduced through vegetative means rather than seed orchards.

The genetic make-up of plants can be changed by the introduction of foreign genes into cells which acquire new characteristics as a result of the introduction. This is genetic engineering which results in a transformed plant. Genetic engineering involves selecting a gene or piece of DNA which contains the gene of interest, packaging it so that can enter the DNA of the new target plant, and then transferring the packaged DNA into the DNA of the target plant.

A prime strategic target for genetic transformation of plantation trees is plant sterility through the total absence of reproductive structures. This strategy will provide the following:

- Selected genotypes with this gene will not be able to reproduce, will be contained within the plantation and will not escape into surrounding native vegetation.
- Energy and nutrients previously directed to reproductive organs will be directed to wood production.

With foreign genes contained through sterility, other useful genes can be introduced into elite plantation trees without any threat to the environment. Genes that will confer insect resistance and herbicide resistance have already been developed (Edwards *et al.* 1995). It may also be possible to engineer trees for resistance to environmental stresses including tolerance to salinity and heavy metals. Candidate genes for stress tolerance are currently under investigation. The genetic engineering approach to developing stress tolerant trees may extend the level of salinity that can be tolerated beyond the level that exists in natural populations. This work will also contribute to our understanding of the mechanism of stress tolerance in plants.

Propagation systems

Genetically improved or superior material requires a suitable propagation system which allows the material to be reproduced on a scale suitable for routine planting. Very high quality seed, such as that from controlled pollination, may only be produced in small quantities, however it can be bulked up using a variety of vegetative systems.

Whether developed through genetic mapping or by genetic engineering, vegetative propagation methods offer the means to rapidly deliver improved tree varieties to the field by bypassing the production constraints of traditional seed based propagation. Vegetative propagation also enables the establishment of a uniform plantation crop, with a predictable level of performance and quality.

Plantations have traditionally been established using seedlings raised in containers or in open-rooted nurseries. For example, *P. radiata* seed is generally produced in seed orchards and rated according to potential growth quality for Australia and New Zealand (Growth and Form rating is often used). Better genetic quality seed is produced in small quantities through controlled pollination and this seed is expensive. Vegetative systems have been developed to amplify plant numbers so a larger area can be planted with superior material. New genetically-modified material will also need amplification and a number of options are available.

Mass propagation

Mass propagation is the integrating technology which enables the gains from breeding systems to be multiplied faithfully and in sufficient numbers for commercial application. Clonal propagation is required to fully capture the benefits of improvement systems using higher levels of technology such as marker aided selection and genetic transformations.

Three types of vegetative propagation methods used for plantation forestry are:

Cuttings

This procedure, which takes shoots from elite parent plants for raising in nursery beds, is simple and reliable. It is an appropriate technology where bulking up of stock plants can occur in parallel with field testing to identify elite trees.

When elite plants developed by genetic mapping or genetic engineering can be identified as seedlings without the need for further field testing, then the multiplication rate achievable with cuttings is generally too low to enable such material to be quickly brought to the field in substantial numbers.

Cuttings are produced in nurseries from stock plants, which are trimmed in spring to allow production of shoots for setting. Cuttings are usually set directly into raised beds or containers in winter and grown in a similar regime to seedlings. In the case of pines this system is very reliable and cost effective, but at this time is much more variable for *Eucalyptus* and *Acacias*.

Somatic embryogenesis

In this procedure, embryos are removed from elite seed and cultured to produce cell suspensions or are grown on a solid medium; the cells are then stimulated to grow into embryos and subsequently germinate into plants. The main problem with this technique is that embryogenic competence is genetically controlled and conversion from embryoids to plantlets is usually low. In *P. radiata* for example, less than 5% of genotypes are competent for somatic embryogenesis; the method is costly, and embryonic cultures are difficult to establish and maintain.

Micropropagation

Propagation from small pieces of plant tissue in an artificial growing medium is now robust for many forestry species. The main drawback of this technique is that it is labour intensive and expensive. This technique lends itself to the application of robotics in order to reduce labour costs, maintain quality and eliminate infection, but at this time the efficiency for forest species is low.

Clonal material of *P. radiata* is routinely used in plantation management. While the value for eucalypts is recognised it is not so regularly used and there are some technical difficulties with some of the species, especially those in the subgenera *Corymbia* and *Monocalyptus*. Clonal and seedling material of *E. camaldulensis* have been successfully tested in Victoria demonstrating faster growth and greater salt resistance than unselected populations (Morris 1995). The work demonstrated significant site interactions. Bell *et al.* (1993) assessed morphology of genotypes (clonal) of *E. camaldulensis*. Clonal 9-month-old plants were less variable in both above- and below-ground morphology than seedling material populations of the same age, this being a critical factor for further development. They concluded that clonal *E. camaldulensis* had advantages in plantations where there were saturated, saline and heavy soil conditions. Progenies for use of selection and micropropagation of *E. camaldulensis* have been established for commercial plantations and in other areas for a wider range of uses (McComb *et al.* 1989).

Tissue culture

Tissue culture by organogenesis or embryogenesis offers alternative systems for multiplication. Embryos from control pollinated seed are usually the starting material for organogenesis. When shoots are large enough they are set as small cuttings in containers to form roots. After rooting, they are lined out in nursery beds and grown on like seedlings or cuttings. Embryogenesis in pines is initiated from immature seed. The process is still being developed but will allow the equivalent of control pollinated seeds at lower cost. The main drawback is the high cost so the material being produced would need to have a high value to justify the process.

Programs involving tissue culture for *E. camaldulensis* improvement programs have been developed. The value of such material is demonstrated for both commercial plantations and analysis of physiological processes (Kabay *et al.* 1986; McComb *et al.* 1989; Bell *et al.* 1993). The principal of development of eucalypts for saline areas is demonstrated but further work is required on a range of species.

Concluding comments

Selecting and improving suitable genetic material for commercial plantations is critical for success. However, this requires a systematic and rigorous approach considering site and other characteristics. Tree improvement usually takes long time periods but a range of procedures can ensure short- and long-term gains.

CHAPTER 5

PRODUCTS, QUALITY AND MARKETING

Issues

Large areas of land are increasingly being affected by salt and this results in decreased productivity. Planting of trees is important for reclamation of these lands, however costs are high. Establishment of commercial plantations is a consideration since generated income offsets the costs. The properties of wood products from such plantations need to be understood and considered in the plantation planning stage. The wood produced may be used for high quality products, reconstituted products, fibre, energy and other secondary products.

Plantation development

Establishing tree plantations on degraded landscapes provides environmental benefits such as lowering of watertable levels, reduction of soil erosion, improvement of soil physical and chemical attributes, and improvement of run-off water quality. Such improvements show up in increased soil organic matter, improved infiltration rates, and higher levels of base saturation, that is increased productive capacity. However, establishment and maintenance of such forest plantations involve significant costs. Hence the development of commercial plantations and sale of the products is potentially a key land management and environmental option to generate income on degraded lands. While such commercial plantations can take various forms there are several elements common to all, namely, the requirement for technical knowledge, resources, markets for products, and a capacity to measure and utilise secondary benefits.

Technical knowledge required for development of commercial plantations involves an understanding of site, establishment and management of the plantations, genetic resources, and the possible products and their value. The ability to appropriately apply this information is of critical importance. The required resources include access to funds, suitable land and genetic material. Secondary benefits from plantations, such as changes in the plantation environment, specifically soils and water, greenhouse gas benefits through accumulation of carbon in organic matter, and effects on adjacent landscapes, need to be understood and quantified. Markets for timber products are the key issue for commercial plantations and establishing such plantations without an understanding of what is required, the properties of the products and where they would be sold, will probably lead to failure. The greater both the intensity of management and the level of environmental stress (for example increased salt), the higher the levels of technical information required to ensure success compared with traditional plantations. That is, a higher level of technical knowledge is required to establish and maintain plantations in saline environments than in more traditional environments. Such an analysis determines characteristics of suitable tree species.

Plantation planning and management

Management of planted forest crops has a different focus to traditional natural forests. Commercial forestry involves a range of plantations from areas totally committed to forest production where the trees are reasonably closely spaced, through to areas with much wider tree spacing and incorporating other activities such as grazing (agroforestry). A number of planting configurations may be used and can include uniformly widely spaced trees or clumps or line plantings on the edges of land used for other purposes. All will have benefits and a number of such options may be utilised within a single project area. The important factors are that the individual trees are planted to provide a specified commercial end product and that sufficient areas of trees are planted on a local or regional basis to support both a market and the associated marketing and transport infrastructure.

Planting of trees for the broad purpose of landcare and catchment protection produces environmental benefits as an end in itself. Commercial plantations need to consider markets and potential changes in them, and develop the plantations to meet such requirements. Such market considerations involve analyses of the product type (sawnwood or veneer, structural wood, pulp or fuelwood), quality and value. Eventually, such an analysis needs to be undertaken for each specific plantation project. However comments of a more general nature can be considered.

Most of the areas established to plantation in the foreseeable future will be on sites with tradition characteristics (that is rain-fed), but there are competitive alternative landuses developing for such sites. Hence, significant areas of plantations will need to be established on sites usually considered to be sub-optimal as a result of natural or man-induced factors. These sites include drier areas, and areas affected by erosion, compaction, salinity, waterlogging and/or pollutants. In considering these areas, it needs to be stressed that commercial plantations have the primary objective of producing a product to be sold for an acceptable financial return. The level of return considered to be acceptable will be determined by the individual landowners or managers and their analyses of alternative landuses, and can be quite variable depending upon the enterprise. In the ensuing analysis, the site amelioration obtained from planting trees is a factor which should be included.

Key areas of expenditure in the development of a commercial plantation include land, management inputs (establishment, weed control, fertilisers), maintenance and harvesting. In terms of outputs, the level of productivity is of critical importance, together with the unit value of the product and the rotation length; this last aspect determining the investment period.

In Australia and many other parts of the world, many of the major plantation programs were initially undertaken or ultimately overtaken by Government agencies, and the objectives of these programs were often basically social or politically driven rather than having primarily a commercial basis (Carron 1990). For many organisations this is currently changing with a direct focus on competitive commercial programs. Such changes are reflected in governments selling established plantations to private enterprises.

Tree species vary in their characteristics as do their products. The specific market and its requirements need to be identified and accommodated in the establishment phase of the project to ensure greater efficiency. Such product and market identification will characterise not only potential species, but the rotation length and spatial scale of the project and will be important where there is a large potential to modify the product through genetic improvement in accordance with market requirements.

Project development

A plantation project may focus on producing a single product, such as pulpwood, or multiple products, such as pulp, roundwood and sawlog, and identification of the product is critically important for the planning process and appropriate management (Box 5.1).

Profitability of any plantation project is directly affected by the length of the rotation as this determines the investment period. Short rotations provide returns from sales in relatively short periods of time, but the unit value of the products ($/m^3 or $/tonne) is generally low. To offset the low product value, the productivity needs to be high and transport distances to market processing need to be short and accessible. Short rotation high productivity fibre crops allow for improvement of the products through breeding and other management procedures to be highly focussed. For example, in producing for a pulp market the most desirable attributes, such as low wood density or low lignin content, can be identified and included in a short turnover period.

Longer rotations shift the emphases to the higher quality markets which yield a much higher dollar value per unit of product. The product unit values need to be high to cover the much longer investment period. Management can be improved through identification

Box 5.1 Attributes of forest plantation projects and their relationship to species and products.

Characteristics of commercial forest products identify alternative suitable species for planting. Selection of species allows for the identification of management and site requirements. Examples of selected species have been given below but the same principles apply over a much wider range of situations.

EXAMPLE 1

Primary target product:	**High quality paper utilising short fibre hardwood**
Secondary product:	No secondary commercial product. Secondary environmental benefits through increase in depth to high watertables and increasing soil organic matter. Increased land value at end of rotation.
* Species:	*Acacia mangium, E. globulus, E. grandis, E. camaldulensis* x *grandis, E. urograndis.*
* Management:	Rotations of 7–10 years with no thinning or pruning.
* Planting stock:	Clonal.
* Productivity:	30–40 m^3/ha/yr mean annual increment.

EXAMPLE 2

Primary target product:	**Newsprint utilising coniferous pulpwood**
Secondary product:	Nil.
* Species:	In temperate areas, radiata pine (*P. radiata*) while in tropical areas, Caribbean pine (*P. caribaea*).
* Management:	Short rotation of 12–15 years with no thinning or pruning.
* Planting stock:	Clonal.
* Productivity:	20–25 m^3/ha/yr mean annual increment.

EXAMPLE 3

Primary target product:	**Sawn softwood construction material**
Secondary product:	Pulpwood, from thinnings.
* Species:	*P. radiata,* a species readily sawn and dried.
* Management:	Rotations of 25–30 years with thinnings for pulp and intermediate small sawlog selection. Pruning undertaken to produce clearwood and applications of fertiliser intra-rotation to increase productivity.
* Planting Stock:	Clonal or seed from control pollination.
* Productivity:	20–30 m^3/ha/yr mean annual increment.

EXAMPLE 4

Primary target product:	**High value cabinet timber (hardwood)**
Secondary product:	None identified.
* Species:	*Acacia melanoxylon*, rainforest species, *E. grandis*
* Management:	Rotation lengths of 40+ years with pruning and selected fertiliser applications.
* Planting stock:	Selected seedlings.
* Productivity:	12–18 m^3/ha/yr.

EXAMPLE 5

Primary target product:	**Thinned, peeled reconstituted products** (Development of veneers and production of laminated veneer lumber or plywood.)
Secondary product:	Woodchips, fuel, and Medium Density Fibreboard.
* Species:	*Eucalyptus grandis, E. urograndis, E. globulus*
* Management:	Rotation length of 15 years with pruning and selected fertiliser applications.
* Planting stock:	Clonal or control pollinated seed orchard material.
* Productivity:	25+ m^3/ha/yr.

and quantification of the attributes which give the product its quality. For example, for sawn structural material, value is improved with, among other things, straight grain, and small diameter knots or no knots. The former can be achieved through pruning while the latter is through tree improvement programs. Characteristics which improve value need to be recognised by grower and buyer and appropriate premiums be paid. Where the quality of the material is its appearance, such as for peeling, those characteristics again need to be recognised. For example, in face material of flooded gum (*E. grandis)*, high value may result from deep red colouration with wavy grain. Such characteristics may not become apparent until trees are relatively mature.

There are interactions between potential products, their value and the spatial and temporal extent of a project and they are largely determined by the required size and upper limits for processing. The quantities of relatively uniform low value wood required for a pulp mill are large (1–2 million tonnes per year) whereas a slicing veneer mill may require small quantities of very high quality timber. While each project needs to be analysed separately, some relationships are outlined in Table 5.1.

Economic considerations

Considerations of the value of projects normally involve discounted costs and returns. These vary according to locations and organisations, however the principles can be generally considered. The use of land for forest plantations needs to be compared with alternative land uses, such as grazing or agriculture. In any projects, the major costs are in the first few years and are

Table 5.1 Product type, species, rotation length, and required areas for plantation projects.

Product quality rating	Product examples & characteristics	Species	Rotation length (years)	Required area* (ha)
Very high	Face veneer (Beauty)	*P. radiata*	30–40	2 500
		E. grandis	30–40	3 000
		E. camaldulensis	40–50	2 500
		E. delegatensis	40–50	3 000
* (Assumes establishment for veneer production only and supporting one mill)				
High	Sawlog: sawn structural material (Strength)	*P. caribaea*	20–30	15 000
		P. radiata	20–30	15 000
		E. grandis	25–40	20 000
* (Assumes support for single medium-sized local sawmill)				
Medium-high	Rotary peeled veneer Laminated veneer lumber, plywood (Consistency)	*E. globulus*	20–30	10 000
		E. grandis	20–30	10 000
		P. radiata	20–30	10 000
Moderate	Woodchips: paper, fibreboards (Structure)	*E. globulus*	10–15	25 000
		E. grandis x *urophylla*	8–12	20 000
		A. mangium	7–10	20 000
		Eucalyptus 'hybrids'	10–12	20 000
Low	Poles and posts (Strength)	*P. radiata*	15	2 000
		P. caribaea	15	2 000
		E. maculata	20	2 500
* (Usually part of a larger, integrated process. Aimed at commercial processes rather than on-farm production. Ability to treat timber with preservatives is important.)				
Low	Firewood (Basic energy content)	*Eucalyptus* spp. *Acacia* spp.	15–20	1000
* (Assumes proximity to market. May be developed for specific energy projects and areas will be modified accordingly.)				
High	Essential oils	*E. polybracta*	3–7	500

associated with land purchase and establishment. For most forest plantation projects there is a period of investment before revenues start to flow. In many areas this period may be 15 years or more, however in the case of fibre crops there will be a return in less than 10 years.

Analysis of investment in establishment of plantations on saline and/or waterlogged areas requires very detailed examination. In such areas land values may be lower, productivity may be reduced and there may be a pattern of continued land quality decline. Planting of trees may have an outcome of reversing such trends. In traditional investment analyses environmental effects are not considered, but there are significant reasons for plantation development in saline and waterlogged areas and they need to be assessed. Reduction in watertables, increased organic matter and stabilisation of salt need to be valued in the economic evaluation either directly or indirectly as changes in land values. While direct economic returns may be lower than with traditional plantations, environmental effects will generally add to making the system viable. Individual landowners alone may not be able to undertake such programs, rather regional implications need coordination.

The analyses can be quite complex when environmental accounting is included. Tree planting in saline or waterlogged areas often compares commercial plantations with landcare or protection types of projects, but such comparisons are artificial. For landcare purposes, the establishment and suitable growth of a range of species is an end in itself, however for commercial plantations, growth and establishment are assumed and analyses need to address critical issues including product quality, pricing and value, rotation length and minimum requirements for plantation areas.

When considering establishment of new plantation resources, the context of the existing and developing resources in the region together with proximity to existing or new processing facilities and ports need to be considered. For example, within Australia there are more than one million hectares of plantations primarily consisting of coniferous species. The coniferous species are dominated by *P. radiata* in more temperate areas and *P. caribaea* and *P. elliottii* in the tropics. There are more than 140 000 ha of hardwood plantations composed of a wide number of species, including *E. globulus, E. regnans, E. nitens, E. pilularis* and *E. grandis.* The total area of softwood plantation is 959 000 ha and the hardwood plantation area is 47 000 ha (McLennan 1996). These plantations are predominantly in traditional rain-fed areas and for some specific reasons (to limit weed infestation or soil erosion); few are in areas with increasing salinity and hence broadscale experience is limited.

Characteristics of forest products

The desirable characteristics of timber depend upon the desired/planned product and they need to be assessed for each individual resource. Broad characteristics of interest include the density of timber, fibre length, and numbers and sizes of knots. Large between-tree variation in wood properties are known to exist and do not generally correlate well with environmental conditions suggesting that a considerable proportion of the variation is due to genetics.

Fundamentally, wood is a structured lignocellulosic material. Being a biological material derived from a number of different species and grown under widely different conditions, there is a large variation in the properties, and the specific property of interest will depend upon its planned end use. When seen as a structural material, the engineering properties of wood are of major concern; when used internally in dwellings, aesthetics are of interest. As a fibre source, physical and chemical properties are of interest, while as an energy source its calorific value and chemical properties need to be known, and where non-wood products (such as essential oils) are used, chemistry is a major concern. A detailed presentation on the properties is not provided here, rather some of the issues specific to plantation development are outlined.

Essential oils

Essential oils are produced by many plants and stored in structures within various organs and tissues such as flowers, fruit, leaves, roots, bark and wood. It is the oils that give eucalypts their characteristic odours. These oils can be obtained from plants by steam distillation and separation. Oils obtained by steam distillation include those of *Eucalyptus*, *Melaleuca*

Table 5.2 Commercial *Eucalyptus* oils (Lassak 1988; Boland *et al.* 1991).

Species	Principle leaf oil constituent	Oil (%)	Oil (%) on fresh weight basis
Medicinal oils			
E. camaldulensis	cineole	10–90	0.3–2.8
E. cneorifolia	cineole	40–90	~2.0
E. dives (cineole variant)	cineole	60–75	3.0–6.0
E. dumosa	cineole	33–70	1.0–2.0
E. goniocalyx	cineole	60–80	1.5–2.5
E. globulus	cineole	60–85	0.7–2.4
E. leucoxylon	cineole	65–75	0.8–2.5
E. oleosa	cineole	45–52	1.0–2.1
E. polybracta	cineole	60–93	0.7–5.0
E. radiata spp. *radiata* (cineole variant)	cineole	65–75	2.5–3.5
E. sideroxylon	cineole	60–75	0.5–2.5
E. smithii	cineole	70–80	1.0–2.2
E. tereticornis	cineole	45	0.9–1.0
E. viridis	cineole	70–80	1.0–1.5
Industrial oils			
E. dives (phellandrene variant)	phellandrene	60–80	1.5–5.0
E. dives (piperitone variant)	piperitone	40–56	3.0–6.5
E. elata (piperitone variant)	piperitone	40–55	
Perfumery and flavouring oils			
E. citriodora (citranellal variant)	citranellal	65–80	0.5–2.0
E. macarthurii (leaf oil)	geranyl acetate	60–70	0.2–1.0
E. macarthurii (bark oil)	geranyl acetate	60–68	0.1–0.4
E. staieriana	citral	16–40	1.2–1.5
E. campanulata	methyl cinnamate	95	1.6–6.1

(tea tree) and *Mentha* (peppermint) (Boland *et al.* 1991). Depending upon the species, the concentration of oil in the leaves ranges from undetectable to 0.1–0.5% on a fresh weight basis. Some of the oils appear to provide protection from leaf eating insects (Morrow *et al.* 1976; Morrow and Fox 1980; Boland *et al.* 1991).

For commercial purposes, the oils are broadly used for medicinal, industrial and perfumery/flavouring. Most medicinal *Eucalyptus* oils have an active therapeutic constituent of 1,8-cineole, and oils are graded and priced on their content of this component. The oils are widely used as inhalants and as an antiseptic.

Industrial oils principally contain piperitone and α-phellandrene. At present α-phellandrene is used for scenting in inexpensive disinfectants and industrial liquid soaps. There are a wide number of other industrial uses. The market potential of *Eucalyptus* oil is based on its solvent properties. The attraction of solvent markets is that they are large, diverse and mainly supplied with a product (trichloroethane) which is being withdrawn from use, under international conventions, to control ozone depletion. There is a strong preference in these markets for 'natural' replacement products. Few *Eucalyptus* species have been exploited for perfumery but citronellal has been used for direct scent use or to develop more valuable perfumes (Lassak 1988).

There is considerable genetic variability in oil content and composition (Table 5.2). Within species, 'chemical forms' have been identified (for example *E. dives* has five chemical forms, such as *E. dives* cineole variety) Johnstone (1984). Within a species, the heritability of oil yields has been shown to be very high, providing great potential for genetic improvement in composition and yield (Shiva *et al.* 1988; Bartle 1995). Such changes could be for direct commercial use or to improve resistance to specific insects. Leaf oil contents and yields reported in Western Australia from trial plots range from a modest 3.5% and 40 kg/ha/yr (Eastham *et al.* 1993) to higher levels (4% and 200 kg/ha/yr). In studies on *E. camaldulensis* in a 3.5-year-old trial of open pollinated material, selected from a broad analysis of tropical eucalypts (Doran and Brophy 1990) from Petford, Queensland, 1,8-cineole levels were found to be highly heritable. As the oil contents were negatively correlated with productivity, both characteristics could be improved simultaneously. First generation increases of 25–32% in oil yield were expected. There was a strong site x genotype interaction for 1,8-cineole content (Doran and Matheson 1994) and this can also be affected by management inputs (Milthorpe *et al.* 1994).

Concluding comments

Establishment of trees in degraded environments can produce both commercial and environmental benefits when appropriately planned. Such planning requires the identification of key products and implementation of suitable management systems to achieve the desired products. General properties of timber can be identified and can be improved through genetic selection. As markets are identified and the plantation develops, there needs to be more detailed analysis and characterisation of product properties.

CHAPTER

EFFECTS ON ENVIRONMENTAL BENEFITS

Issues

Establishment of fast growing plantations can result in environmental benefits such as modification of soil and water salinity levels, lowering of watertables, sequestration of carbon, and improvement of soil nutrient levels and organic matter. Environmental accounting systems for plantation projects will need to assess and appropriately value any such changes.

The establishment of productive forest plantations has direct and readily measurable commercial benefits in the form of timber and other products. Additionally, there are environmental benefits which are less easily quantifiable on an economic basis. Management of forest plantations in traditional areas recognises that such benefits may be accrued but they are not of primary consideration in plantation evaluation. However, in saline and waterlogged environments, the environmental benefits are a critical consideration in plantation establishment and there is a need to both understand and include such effects in valuing the project. This may take the form of parallel economic and environmental accounting, or the environmental benefits may be directly valued and included in the economic analysis. One example of such benefits is an improvement or change in land values as a result of tree planting. Either by using direct economic analyses or using other methods to account for environmental benefits, the issues related to environmental change are critically important and are outlined below.

Scale of plantations

Planting of trees can result in improvement of environmental values through effects on the hydrological cycle, accumulation of carbon, and the utilisation and redistribution of chemical elements within the ecosystem. Such effects are scale-related within the landscape but actual quantification of effects is often difficult as the information base is limited. Much of the research on forest trees in this regard has been carried out under laboratory conditions using small plants, or in the field on several hectares, or using a proportion of a small catchment, say less than 100 ha. However in commercial forestry the scale and concentration of plantings are much larger. Commercial forest plantations require hundreds or thousands of hectares to be planted annually with a minimum of 10 000 to 20 000 ha on a project area. Hence research predictions require substantial 'scaling up'. Whether plantations are in large blocks, or in sections representing a proportion of a large catchment (say 20%), there will be substantial impacts. The results of increased productivity on such an area are far reaching and factors such as hydrology (for example watertable depth and recharge of soil water) will be significantly affected. The form of planting will be a factor in determining the end result, but to obtain a result at a regional level there is a requirement that significant areas are planted to test the results at that level.

Establishing forest plantations within an area which currently has few or no trees will have both significant short- and long-term impacts on the environment. Where the site is generally saline, waterlogged or creating stress for plant growth in some other way, the effects of plantations may often be ameliorative (Box 6.1). While these impacts will primarily result in modifications to the hydrological cycle and changes in soil properties, other broader considerations such as potential 'greenhouse' gas effects should be considered. Many of the effects have not been systematically quantified but patterns may be established through development of appropriate models.

Hydrological effects

Different types of vegetation use different quantities of water in any period of time and this directly affects the hydrological balance. The production of organic matter in any period of time is a major determinant of water use, although there may be significant variations in rainfall interception. The spatial and temporal scales of analysis directly affect results and

Box 6.1 Processes by which trees may maintain or improve soils (from Ward 1991; Prinsley 1992).

Processes which may augment additions to the soil:

- Maintenance or increase of soil organic matter through carbon fixation in photosynthesis and its transfer via litter and root decay.
- Nitrogen fixation by some leguminous and a few non-leguminous trees.
- Uptake of nutrients released by rock weathering in deeper layers of the soil.
- Favourable conditions provided by trees for inputs of nutrients by rainfall and dust, including via throughfall and stemflow.
- Exudation of growth-promoting substances by the rhizosphere.

Processes which may reduce losses from the soil:

- Protection from erosion and thereby from loss of organic matter and nutrients.
- Trapping and recycling nutrients which would otherwise be lost by leaching including through the action of mycorrhizal systems associated with sloughing of tree roots through root exudation.
- Reduction of the rate of organic matter decomposition by shading.

Processes which may affect soil physical conditions:

- Maintenance or improvement of soil physical properties (structure, aeration, porosity, moisture retention capacity and permeability) through a combination of maintenance of organic matter and effects of roots.
- Breaking up of compact or indurated layers by roots.
- Modification of extremes of soil temperature through combination of shading by canopy and litter cover.
- Profile drying through water recovery from depth (lowering watertable).
- Reduction of salt in the landscape
- Redistribution of salt in the landscape.

Processes which may affect soil chemical conditions:

- Reduction of acidity, through addition of bases in tree litter.
- Reduction of salinity or sodicity

Soil biological processes and effects:

- Production of a range of different qualities of plant litter through supply of a mixture of woody and herbaceous material, including root residues.

Timing of nutrient release:

- The potential to control litter decay through selection of tree species and management of pruning and thereby to synchronise nutrient release from litter decay with requirements of plants for nutrient uptake.

Effects on soil fauna.

conclusions and these are discussed below. If replenishment of soil water in the upper profile by rain or irrigation is insufficient to meet the atmospheric demand for evapotranspiration, tree roots tend to extend more deeply until they are able to access the saturation zone in the vicinity of the watertable or growth is reduced. While many studies have been carried out on water usage by individual trees (Carbon *et al.* 1981; Chaturvedi *et al.* 1984; Morris and Wehner 1987; Raper 1998), the question needs to be addressed on a much broader area in the case of commercial plantations, preferably at the catchment level.

Different effects have been noted and these relate to hydrological characteristics and type of planting systems. Hence patterns of tree planting have different interactive hydrological effects. Alternative plantings within marginal to dry areas are:

1 **Total area plantings.** A large proportion of a catchment area is planted with no specific pattern other than no disturbance on very steep areas and protection of creek lines. This will have major broadscale effects on ground water recharge and run-off.

2 **Recharge areas.** Planting in recharge areas within the landscape reduces quantities of water further down the landscape and will be designed for specific topographical configurations. Only a relatively small proportion of the total area may be planted (less than 15%) to achieve the objectives.

3 **Break of slope plantings.** Water inputs accumulate on slopes and move laterally down the slope affecting levels of water in lower slopes (these are specific instances of recharge area plantings). Plantings at the break of slope utilise up slope water and affect lower slope productivity. Such plantings are in distinctive areas of the landscape rather than broadscale. Trees are using incipient rainfall plus laterally moving water, and hence the hydrological effects will be lower down the slope. A relatively small proportion of the total area is planted.

4 **Plantings in seeps and discharge areas.** Such plantings are in specifically defined problem areas such as up slope of, or in, salt seeps. The aim is to utilise excess water to reduce local and down slope effects, and as such the plantings are limited in area. Most discharge areas within the landscape reduce quantities of water further down the landscape and can be quite broadscale.

5 **Riparian zone protection.** Plantings are established adjacent to creek lines. A primary objective is maintenance or improvement of stream and aquatic values in conjunction with wood production. These areas have a disproportionally large effect on run-off and water quality so impacts can be quite high (Dye and Poulter 1995).

6 **Ground water utilisation plantings.** Such plantations utilise ground water for a significant component of plant water requirement. They may require irrigation during early stages of development but over most of the rotation ground water is accessed and there will be a lowering of ground water, at least beneath the planted area. Density of planting, level of productivity, and the planting pattern will all have an effect on the degree to which the watertable is lowered.

7 **Irrigation of plantations.** In areas with limited rainfall, growth will be improved by irrigation. Environmental effects will occur due to irrigation, while the source of water will also be of significance. Where ground water is pumped from the same area as the plantation, the irrigation can result in the ground water being lowered to a level beyond which it can be directly utilised.

1

2

1 Natural regrowth of Murray River red gum (*E. camaldulensis*), a species primarily growing under riverine situations. One of the forest types which are critical components of the water balance in the Murray Darling area.

2 Dry woodland type (*Eucalyptus spp.*) important for maintaining hydrologic balance.

3

4

5

3 Salt-affected landscape. The area previously supported woodland which has been cleared.

4 Degradation due to rising water table and salinity. The area was partially cleared by converting perennial woodland to annual pasture species which are now declining.

5 Salt crystals on soil surface.

6

7

8

6 Eucalypt foliage showing salt damage.

7 *E. globulus* foliage showing tip dieback due to elevated salt content.

8 Site preparation for plantation establishment in an area with an elevated water table.

9

10

11

9 Cuttings of *P. radiata* established on an improved pasture site.

10 Eucalypt plantation established in a saline discharge area.

11 Farm forestry project with trees planted on a discharge area.

12

13

14

12 Ten years of dryland salinity management theory. **The Engineers' solution:** The far ridge is planted to indigenous varieties (*E. polyanthemos* and *E. macrorrhyncha*) at 25 stems/ha with the objective of capturing rainfall on upper slopes. **The Forester's solution:** The foot of the far ridge is planted to 1300 stems/ha bands of *E. globulus* and *P. radiata* with the intention of capturing subsurface recharge flows at the 'break of slope'.

13 Blue gums (*E. globulus*) established on an old pasture site indicating the components of previous annual vegetation, soil type, site preparation and plantation species.

14 *E. globulus* (left) and *E. grandis* (right) at 9 months of age in a sustainability trial in the Shepparton area, Victoria.

15 

16 

17 

15 *E. sideroxylon* (10-years-old) on a sedimentary ridge in the Warrenboyne area, SE of Benalla, Victoria. The trees were planted as part of a catchment salinity management scheme.

16 A site where 8-year-old *E. globulus* belts are suffering drought losses. These sites have proven unsatisfactory for intensive plantations.

17 *E. grandis* (6-years-old) irrigated with municipal waste water, Goulburn Valley Region Water Authority, Shepparton, Victoria.

18

19

20

21

18 A young blackbutt (*E. pilularis*) plantation (8-years-old) treated with fertiliser and good weed control at planting.

19 A 20-year-old blackbutt (*E. pilularis*) plantation established in an equivalent manner to 18 above.

20 Singled *E. globulus* (approx. 7-months-old) coppice (foreground) with *E. globulus* (3.5-years-old) in the background. The plantations were irrigated with municipal waste water, Goulburn Valley Region Water Authority, Shepparton, Victoria.

21 Developing *E. globulus* plantation (2-years-old) in an area with rising water table.

22

23

24

22 *E. elata* (1-year-old) planted on a degraded site. The trees on the left were fertilised at planting; trees on the right were unfertilised.

23 Eucalypt plantation in an area with rising water table and salt damage.

24 V-notched weir and system for measuring water flow in a research catchment study.

Acknowledgements

Nos. 9, 12, 14, 15, 17 Courtesy: D. Stackpole, Department Natural Resources & Environment. **Nos. 7, 16, 20** Courtesy: T. Baker, Department Natural Resources & Environment. **Nos. 10, 11** Courtesy: J. Collopy, Department Natural Resources & Environment. **Nos. 3, 5** Image provided by the Department of Primary Industries, Queensland from their book, House *et al.* (1998). 'Selecting trees for the rehabilitation of saline sites in south-east Queensland'. Technical Paper No. 52 published by DPI, Queensland. **Nos. 1, 2, 4, 6, 8, 13, 18, 19, 21, 22, 23, 24** Supplied by the authors.

Box 6.2 Water balance in forest plantation systems.

Studies of water use by forests often utilise a water balance analysis which quantitatively sums the inputs, outputs and storage of water. This is often presented in the form of an equation:

$$P_o = ET + I + R \pm \Delta S$$

where:

P_o is the amount of rainfall in an area (mm/yr)

ET is evapotranspiration or the amount of water both evaporated from the soil surface and transpired by the plants (mm/yr)

I is the amount of water intercepted by plant foliage and lost from that surface by evaporation (mm/yr)

R is the amount of water lost by running off the ground surface and by drainage through the soil (mm/yr)

ΔS is the change in water content in the root zone (mm/yr)

The individual components can be further subdivided, usually for estimation purposes for example:

$$P_o = TF + SF + I$$

where the additional factors of throughfall (TF, mm/yr) which is the water passing through the canopy, and stemflow (SF, mm/yr) the water running down the stem are included. By knowing precipitation (P_o), throughfall (TF) and stemflow (SF), the amount of water intercepted by a tree canopy and evaporated (I, mm/yr) can be estimated.

As an example, a 9-year-old plantation of *E. camaldulensis* in Israel was compared with adjacent grassland.

	Eucalypt plantation	Grassland
Precipitation (P_o, mm)	640	640
Canopy interception (I, mm)	96	0
Evapotranspiration (ET, mm)	466	322
Run-off or drainage (R, mm)	83	320
Change in soil water (ΔS)	–5	–2

Karschon and Heth (1967).

The results demonstrated reduction in run-off and a drying out of the soil due to increased water use by the plantation.

Components of the water balance are usually estimated in closed systems, such as small catchments where input of water is only by precipitation. Essentially, it subdivides the precipitation and is often used as a way of indirectly calculating evapotranspiration. A problem arises when an additional source of water for plant usage is by way of ground water and irrigation. However by adding these additional input factors, the effects of plantations on components of the water balance can be monitored.

The water balance for a plantation, open woodland and grassland in an area with access to a deeper watertable can be analysed. The plantation and woodland produced 20 tonne/ha/yr and 3.6 tonne/ha/yr respectively.

	Forest plantation *Eucalyptus*	Natural woodland *Eucalyptus*	Grass crop
Precipitation, P_0 (mm)	590	590	590
Throughfall, *TF* (mm)	476	584	0
Stemflow, *SF* (mm)	26	13	0
Interception, *I* (mm)	88	63	22
Reaching ground, *TF* + *SF* (mm)	502	527	568
Evaporation (mm) (from ground surface)	20	90	180
Transpiration (mm)	905	310	170
Soil change in root zone (mm)	–5	127	218
Deficit (mm) (from deep profile)	418	0	0

The rapidly growing plantation acquired approximately 418 mm of water from deep in the profile, while under woodland and grassland there was recharge in the root zone.

Traditionally, the hydrological cycle is assessed in terms of a water balance equation which balances inputs, outputs and storage of water (Box 6.2). In this case, the use of water through evapotranspiration varies with age and needs to be analysed separately, that is the productivity per unit area is a critical component. The water balance equation in forests has primarily been developed from situations in which inputs and outputs can be readily measured. The potential to use a significant source of ground water requires that inputs to traditional hydrological models are modified and this has been approached by several researchers (Hingston and Galbraith 1998; Hingston *et al.* 1998). Part of the approach has been to directly measure water use by trees rather than deriving indirect estimates through calculation. Details of this have been reviewed a number of times (for example Raper 1998) covering water balance approaches, ventilated chambers, heat pulse, heat balance, meteorological techniques and isotopic compaction of water. Many of the studies are inappropriate for small plots of trees but a number of alternatives have been used.

Ground water utilisation

Forest plantations can utilise significant quantities of water and lead to a reduction in ground water levels. Such an effect has been demonstrated in a number of locations and to varying degrees. Analyses of such results are in part dependent on the scale being considered since different scales raise different issues. Critical considerations move from an emphasis on individual trees and hydrological processes, through to stands and to regional land use planning. The level of scale affects whether impacts can be as a result of the effect of an individual land manager or are a result of regional planning and policy development (Table 6.1).

Clearing of eucalypt forests and woodlands causes ground waters to rise, and has led to an increase in the outflow of water and salt, producing rapid increases in the area of saltland

Table 6.1 Relationships between scale and effects of tree planting.

Scale	Questions	Comments
Plot	Researcher	Often established for research purposes. Genotype survival and growth. Physiological analysis of plot.
Paddock	Individual land manager	Topographical effects. Water use and availability. Spacing of trees, location and configuration in landscape.
Small catchment	Land managers	Water use and availability. Proportion of catchment planted.
Region	Regional planners/ Policy makers	Which catchments contribute the most salt. Where salt is in relation to ground water. How to classify catchments. Concentration of plantings.

(for example in Western Australia, Williamson *et al.* 1987). In terms of overall water balance, the issues are primarily how much water a tree crop uses, and how the water use is related to tree productivity, to the component of water use derived from ground water sources (watertable, aquifers), and to water derived from lateral movement, such as up slope recharge areas. The relative importance of these factors within a catchment will determine the probable overall impact. In relatively dry environments and where plantations have reasonably high productivity, there has been significant utilisation of ground water measured as a lowering of watertable depth. Utilisation of water from up slope may have less apparent effects immediately adjacent to trees but will lead to changes in water contents on lower slopes. The main consideration is to understand the hydrological relationships and constraints in small catchments and plan within them.

The effect of plantations on ground water will vary with the level of incident rainfall. In higher rainfall areas water will primarily be supplied directly from rainfall but in lower rainfall areas either there are declines in growth or there will be demands on ground water. In the higher rainfall areas (approximately 1000 mm/yr) of the western fringe of south-western Australia, clearing was found to cause an immediate decrease in canopy interception (equivalent to 13% increase in available water) and a 10% reduction in evapotranspiration. Clearing native vegetation was found to cause an increase in ground water recharge (equivalent to 10% of annual rainfall). A rise in the watertable of the order of 0.9 m to 2.6 m was found in cleared catchments while in adjacent uncleared catchments, the watertable remained constant or was lowered (Peck and Williamson 1987; Ruprecht and Schofield 1989; George 1990). In lower rainfall areas of less than 720 mm/yr, establishment of eucalypt plantations led to a fall of 4.4 m in the watertable across sites in a nine-year period. Rates of rise or fall of watertables were directly proportional to the degree of reafforestation or clearing. However, it was noted that it required only a small percentage of the catchment to be cleared (< 20%) to cause the watertable to rise, but large areas need to be reforested (30–50%) to cause the water levels to fall (Loh 1985; Peck and Williamson 1987; Anon 1989; Schofield *et al.* 1989; George 1990; Fitzpatrick 1994).

Significantly reduced ground water levels have been measured under tree plantations established on discharge areas. One of the more effective strategies for lowering the watertable in areas with greater than 700 mm annual rainfall in south-west Australia, was to plant dense plantations of trees covering more than 50% of the previously cleared farmland. For example, ground water beneath the trees was lowered by 2.8 m over 9 years. In addition, ground water salinity decreased on all the higher rainfall (>700 mm) sites planted with trees. Salt concentrations in the surface soil also decreased after planting trees on a site with 450 mm rainfall per year, although in some areas salt concentrations started to increase after a period of time. It could not be expected that plantations in discharge areas, with a continued flow of saline ground water to the trees, will maintain health and be able to control excess water in such a hydrologic setting in the long term. Further modifications such as plantings in recharge areas would also be needed. The area of reforestation required to control salinity within a catchment is uncertain. Estimates for higher rainfall areas range from 20–40% or more (Peck 1978; Williamson 1978; Williams 1979; Engel 1987; Schofield 1989; 1991a, 1991b; Ward 1991). Plantations located on discharge areas are a short- to medium-term component of an overall longer term salinity control strategy. The primary method to control excess water in the longer term is to plant trees in the recharge areas and control all water on that site (Stolte *et al.* 1997).

Plantations are successful where trees are planted in appropriate topographical positions. In catchments with lower rainfall in south-west Western Australia (300–600 mm/yr), efforts have been made to define small portions of the landscape where reforestation could be used to limit recharge and thus salinity (Smith 1962; Bettenay *et al.* 1964; Nulsen and Baxter 1982; Dyson 1983; Clifton 1992; Barker *et al.* 1995). There has been wide variability in recharge zones and this has resulted in the planning for commercial forestry being more difficult. The inability to control ground water recharge, in terms of quantities, locations and responsible processes, led to trial plantings of eucalypts on or adjacent to discharge areas (saline seeps). The watertable has been significantly lowered in some areas (Bell *et al.* 1988). However in the study areas where rainfall was > 700 mm/yr, 30–50% of the catchment needed to be planted in order to significantly lower the watertable in the region over a ten-year period. While some concerns still exist with regard to the ability of plantations to transpire saline ground water in the longer term, studies by Sonogan and Patto (1985), Engel and Negus (1988), and others have indicated plantations have survived and been productive even with reasonably high saline ground waters. Schofield *et al.* (1989) indicated that ground water quality had been improved due to plantation establishment.

In terms of the hydrological balance, planting of commercial crops requires a minimum amount of water for acceptable productivity. In traditional rain-fed forest areas, productivity is affected by the total quantity of rainfall, its seasonality and the ability of soils to store water. Models relating these factors have been developed on a broad scale for a number of plantation species (Czarnowski *et al.* 1971; Inions 1991; McGrath *et al.* 1991; Weston 1991). Such acceptable rainfall is usually above 800 mm with some situations being below 700 mm. In saline areas rainfall may be as high as 800 mm, but many plantations are being considered in the 400–600 mm rainfall range. If productivity is to be maintained, additional sources of water have to include irrigation, influx laterally, or deep storage in the soil. The use of this water by the plantation will be indicated by some depletion in water storage, which will be a direct function of the excess plantation productivity compared with water use from incident rainfall. That is, natural rainfall supports a particular level of productivity

Table 6.2 Reduction in water storage (mm) in the 0–27.5 cm soil profile in the period from January to May.

	(Stems/ha)			
Species	**25**	**75**	**150**	**250**
P. radiata	55	41	72	89
E. globulus	39	72	76	86
E. camaldulensis	16	81	218	200
E. ovata	46	90	57	160
C. cunninghamiana	42	52	84	110

and any additional productivity will be by using irrigation or utilisation of soil water. The greater the unit area productivity, the greater will be the water usage. In Victoria, Bird *et al.* (1993) reported the effects of number of stems per hectare and different species on water use by 9-year-old trees (Table 6.2). Actual productivity such as volume (kg/ha/yr) was not reported and stocking was low (25, 75, 150, 250 stems/ha).

Long-term water use in a plantation of river red gum (*E. camaldulensis)* was estimated using a ground water hydrographic separation technique compared with the transpiration rate estimated by sap flow measurements using the heat pulse method (Salama *et al.* 1994). Recession components of hydrographs of wells inside and outside the plantation were compared for 24 months. During the summer, the difference in gradient between the two wells was assumed to be plantation water use. In summer, water use was 0.9 to 1.33 mm/day and was 0.95 mm/day in winter. Stand transpiration by heat pulse was 0.76 mm/day. Native woodland was found to be 0.9 mm/day. For *E. camaldulensis* seedlots planted under saline and high watertable conditions where the effective rainfall was 950 mm, the best seedlots transpired 1882 mm over 708 days between ages 2 and 3.5 years (average of 2.6 mm/day). The difference of 932 mm between transpiration and rainfall explained the lowering of the watertable by 1.2 m (Zohar and Schiller 1998).

Simple theoretical calculations can be made on water usage by different productivities using water use efficiency factors, that is using a factor which indicates how much water is used to produce a unit of organic matter (measured as net productivity) (Turner *et al.* 1992). Increased use of water by more productive stands and the consequent effects on factors such as soil drying have been reported in a number of studies, such as for *E. grandis* (Boden 1990). In this case, Boden was concerned that soil water content had been depleted so far that growth was affected. Such productivity, as indicated in previous chapters, will be affected by the inherent productivity of the species, the number of trees that are planted per unit area and the management inputs to improve productivity (fertilisers, site preparation). Schematically, the excess water required by a stand (water in excess of natural rainfall) can be considered in terms of the level of rainfall and the productivity of the stand (Fig. 6.1). At low productivity levels (Productivity Class 5) the stand is not using all the incident rainfall, probably indicating other factors, such as lack of nutrients, are limiting growth. As productivity is increased, a point is reached where all rainfall is utilised and excess demand is indicated as a negative estimate. In practice, where ground water is available or irrigation is supplied, productivity will be maintained; and if not, productivity will not be realised.

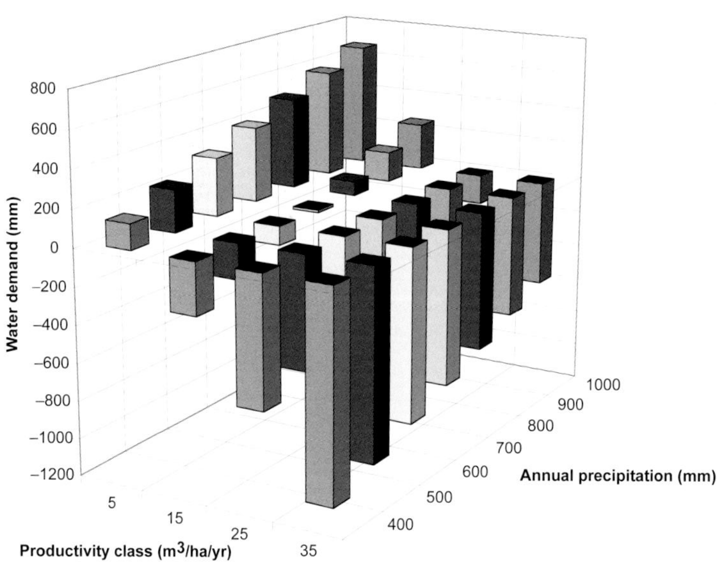

Figure 6.1 Pattern of plantation water demand related to mean stand productivity and average annual rainfall (mm). The positive values indicate that the plantation is not utilising all available rainfall whereas the negative values indicate a deficit to be filled by ground water, by irrigation, or alternatively, the growth potential is not met.

Water demand may be translated into a pattern of ground water declines, recognising the values are only indicators (Fig. 6.2). These estimations are schematic and have used mean annual productivities rather than the more realistic current annual increments. Variation with time, particularly in relation to increasing tree maturity, has not been addressed. The data in Figures 6.1 and 6.2 provide an explanatory framework in which to fit some of the field studies, especially considering the range of results that have been found. The considerations in assessing such work include some estimate of productivity and the rainfall regime together with some estimate of impact on the hydrological cycle. High rainfall is usually associated with lower evaporation. The same volume of water supplied as irrigation to flatter, more arid, warmer, summer areas may lead to more growth and more water use.

Variation in water usage by plantations will occur as the plantation ages with the greatest demand for water occurring near the time of stand crown closure. The actual amount of water that is used is determined by the productivity of the stand and this changes with increasing stand maturity (Fig. 6.3). In normal rain-fed plantations, periodic drought will affect the annual productivity, however there will be increased variability effects in forests dependent on ground water. The systems outlined are not functional models and do not attempt to address the feedback mechanisms which are critical to the maintenance of the plantation productivity in the long term.

Plantations established for ground water discharge control should be located up-catchment from discharge areas, where the ground water is accessible but concentration of salts due to surface evaporation is minimal. For application to trees, saline ground water should be mixed or if possible alternated with fresh water irrigation to maintain a vigorous canopy capable of maximum water use.

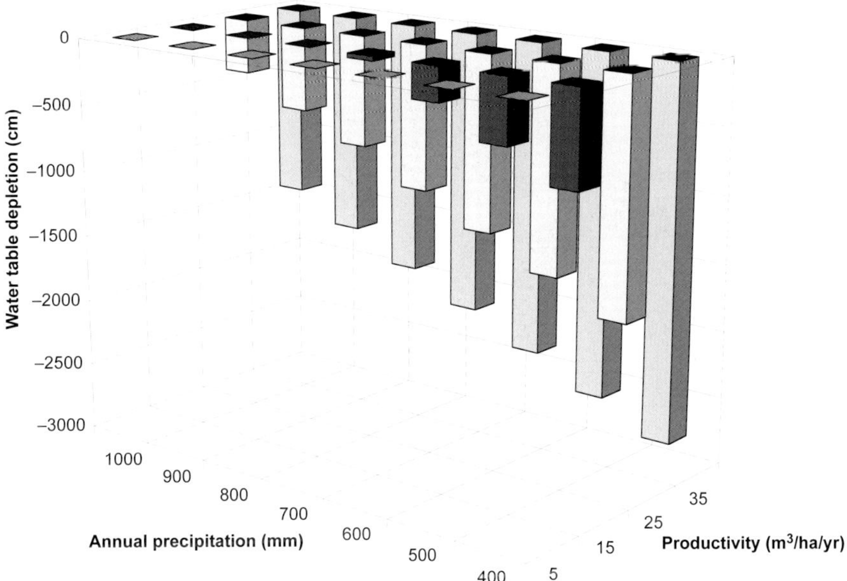

Figure 6.2 Pattern of ground water decline related to plantation growth and average annual rainfall.

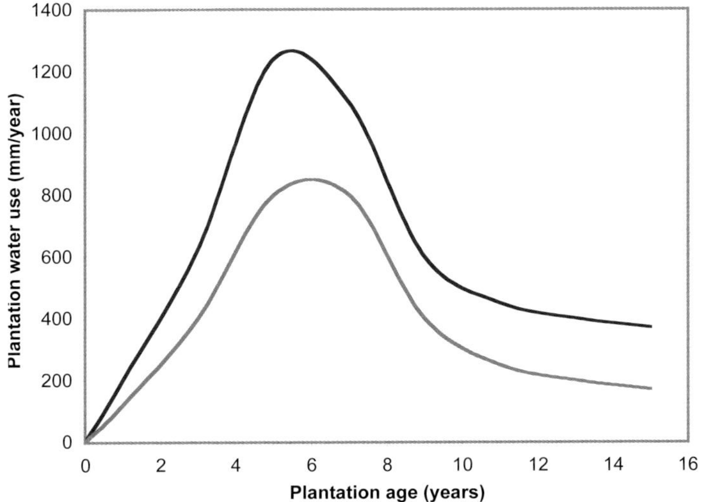

Figure 6.3 Change in water use by two *E. camaldulensis* plantations with different productivities and increasing maturity (based on J. Turner *unpubl. data,* Ranasinghe and Mayhead 1991).

The use of trees to modify ground water conditions has produced variable results and has been questioned as a means of controlling ground water discharge areas and saline seep control (Jenkin 1981; Greenwood 1986; 1992, George 1990). It has been considered that management should initially emphasise the recharge areas (Morris and Thomson 1983; Greenwood 1986).

The establishment of trees and shrubs is a good long-term option for controlling dryland salinity. Hydrogeological features of a catchment both control and cause salinity, so the establishment of trees needs to include the mapped hydrogeomorphic features. Catchment planning should ensure that trees are suited to the area, are preferably of commercial value, and planted in the best hydrogeologically determined positions at sufficient density to reduce excess recharge. Other features will include other biological methods to increase water use within the catchment, improved crop and soil management, and the introduction of salt-tolerant plants. Such systems of planning have been developed and demonstrated (Farrington and Salama 1996).

Selected case studies

Case studies provide an understanding of interacting biological, site and management factors. A potential option for watertable control in shallow watertable areas is the establishment of tree plantations as there are returns to the grower through wood production and improved watertable conditions.

Kyabram Plantation, Victoria

The study area is located at Kyabram in Central Victoria, Australia. It has total of 2.4 ha of *E. botryoides* (southern mahogany), *E. grandis* (flooded gum), *E. camaldulensis* (river red gum), *E. globulus* (southern blue gum) and *E. saligna* (Sydney blue gum). Rainfall is estimated to be about 494 mm/yr and mean pan evaporation is 1345 mm. The soil is a red-brown duplex soil. Trees were irrigated for 6 years. Hydraulic conductivity ranged from a mean of 1.77 cm/h to 0.1 cm/h at 350 cm depth. The study which included modelling techniques had two components, namely field experimentation and model development. The effects of ground water on pasture and plantation were studied (Feikema and Connell 1995). There was recharge of the ground water beneath the irrigated pasture and also discharge (loss of water) from beneath the plantation. The loss from the watertable (located at up to 6 m depth) by the eucalypt plantation was 689 mm/yr in dry years and 215 mm/yr in wet years, that is, in dry years, 74% of the water requirement of the plantation was from the watertable and in wet years it was 24%. The net effect was that the watertable was reduced by an average of 2 m below the plantation and the effect at 90 m from the edge of the plantation was 0.12 m. At issue was the fact that piezometric pressure at 10 m depth was 0.63 m higher than the watertable, indicating that discharge was occurring, conducive to salt accumulation in the root zones of trees. In a sandier soil, the drawdown was 0.75 m with the piezometer levels at 0.21 m. That is, the use of water by trees leads to hydrological differentials and consequent water movement, the rate of which is dependent on soil type.

A number of points were noted:

- Watertable levels under trees showed a seasonal pattern, rising during winter and falling during summer, a trend more marked under trees than in surrounding irrigated land. The watertable was gradually rising beneath the older plantation due to trees using progressively less ground water (as growth declined) and/or from increased irrigation outside the plantation.
- There was an inward lateral and an upward vertical hydraulic gradient respectively around and under the plantation.
- Solute accumulation due to concentrated water extraction occurred in the root zone, leading to salt bulges near the watertable associated with maximum root density. The zones of salt accumulation appeared to be dynamic.

- Trees were using more water than provided by rainfall, indicating they were using ground water to meet requirements.
- The trees at Kyabram appeared to be moisture limited.
- There was a dry zone within the root zone of the trees that was subject to little or no leaching by infiltrating rain.
- Tree growth rates were relatively low for 20-year-old trees.
- Limitations to growth at Kyabram were primarily a result of low soil permeability and high density planting (Connell *et al.* 1997).

The accumulation of salt in the rooting zone has raised questions of long-term productivity. The studies on effects of ground water uptake and diffusion on salt accumulation estimated that salt accumulation eventually produced an equilibrium situation which would take between several years and several decades depending on conditions. Short rotations may not reach equilibrium. Further study is required as to where the salt moves. This may lead to further concentration of salt in ground water; it has not been measured (J. Morris *pers. comm.*).

The study concluded that when compared with surrounding watertable levels, the watertable had been lowered by trees. The spatial extent of this watertable control is limited to 40 m from the plantation boundary. This was considered to be due to the low hydraulic conductivity of the soil.

Hotham Valley, Western Australia

The area in the Hotham Valley related to most of a 27 ha naturally forested catchment in Western Australia which was cleared for agriculture in 1966. The area was in the 760 mm rainfall zone (Biddiscombe *et al.* 1981, 1985a; Greenwood *et al.* 1985, 1992). A saline seep had developed mid-slope and in order to reclaim it, 11% of the catchment was replanted to two multi-species eucalypt plantations in 1976, within and above the seep. The plantation established up slope did not lower the watertable even though evaporation from it exceeded that from the pasture crop by 1600 mm/yr. The plantation in and above the seep lowered the watertable by 0.5 m. There were changes in the flow of chloride and also the concentrations of chloride increased under the plantation areas. The changes allowed salt-sensitive species to re-establish in the area of the seep.

Evaporation from the forest stand after seven years was about 2500 mm/yr whereas the pasture crop evaporated about 450 mm/yr. By the ninth year at least, the plantations were permanently phreatophytic on the deep aquifer, except in the seepage area where they were only temporarily phreatophytic on the seasonal perched aquifer. Pasture roots extended only to 0.8 m and evaporation was only a small proportion of the plantation. Evaporation was estimated as:

Evaporation (mm) = 1394 + 294 Leaf Area Index ($R^2 = 0.84$) and hence the developing plantations will lead to rapid increases in water use.

It was demonstrated that roots of *E. globulus* and *E. cladocalyx* extended into the unconfined aquifer. The effects were to increase interception and associated evaporation, and increase water use by vegetation. The absolute effect depended on the location of the trees (Table 6.3). The soil dried out and reduced the leaching of salt, so this tended to result in salt accumulation.

There was large variation between species in growth and water use and effects on the catchment would have been greater with more appropriate tree species selection. Annual

Table 6.3 Effect of vegetation on change in catchment soil water storage (ΔS) between the wettest (September 1983) and the driest (May 1984) months (Greenwood *et al.* 1992).

		Mean water table depth (m)	Mean ΔS (mm)	Area (%)	Catchment ΔS (mm)
Pasture-crop	Up slope	8.5	–7	44	–3
	Mid-slope	5.0	60	29	18
	Total				15
Plantations	Up slope	7.4	–351	6	–21
	Mid-slope	6.0	–57	5	–3
	Total			11	–24
Natural *E. marginata*	Up slope	8.2	–237	16	–38
Catchment				100	–48

evapotranspiration (the sum of interception, evaporation and transpiration) ranged from 1600 to 2700 mm (Table 6.4). When related to the annual rainfall of 680 mm, the pastures allowed 290 mm of recharge, whereas the eucalypts used an additional 1000 to 2400 mm. Notably, salt tolerance is not the primary issue as growth and water use are of greater relevance.

In the multi-species plantation described by Greenwood *et al.* (1992), the catchment had a permanent deep aquifer confined to its lower third and a shallow seasonal aquifer. Larger stores of salt were found down slope in the soil profile than mid-slope. Chloride in the deep aquifer fell by 20% in seven years, being greater mid-slope. Tree growth was most rapid in the first five years with higher water usage by the trees than the adjacent crop. The plantation would have been more successful in controlling salinity if there had been more careful plantation location (see Fig. 6.4 and Table 6.4).

Table 6.4 Annual evapotranspiration associated with various eucalypts and pasture (Greenwood *et al.* 1985).

		Annual evapotranspiration (mm)
Pasture		390
Up slope planting	*E. maculata*	2300
	E. globulus	2700
	E. cladocalyx	2700
Mid-slope planting	*E. wandoo*	1600
	E. leucoxylon	1800
	E. globulus	2200

Stene's Farm, Western Australia

Stene's Farm is south-east of Perth, Western Australia with annual rainfall of 713 mm and evaporation of 1600 mm. The area was cleared of native forest in the 1950s. The cleared valley floor was planted to trees in 1979 and a strip of grazing land retained as control. The

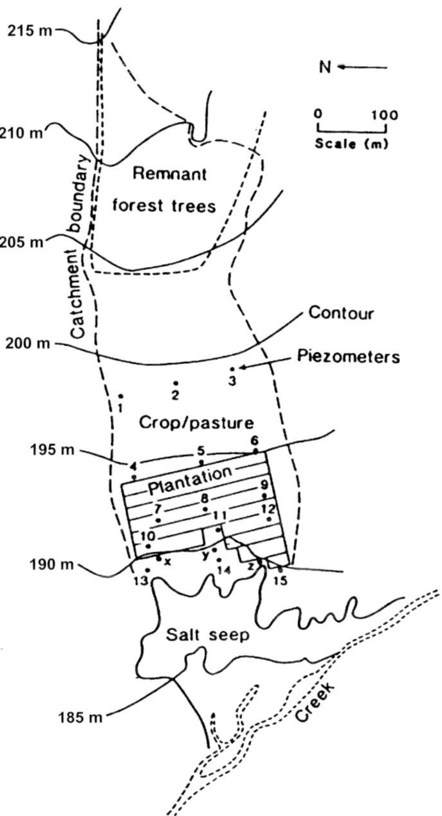

Figure 6.4 Catchment topography and location of vegetation and peizometers.

soils are silty in texture. At time of planting to trees, the ground water with salinity of 5300 mg TSS/L had risen to within 0.5 m of the soil surface.

The eucalypt reforestation was successful in lowering the ground water by 1.5 m while ground water beneath adjacent pastures had risen (Fig. 6.5). The ground water salinity declined by 30% over the ten-year period (Bari and Schofield 1991). Reforestation in areas of the farm was successful in substantially lowering the saline ground water across the site. Between 1980 and 1989, the average minimum ground water levels beneath the trees declined by 5.5 m relative to the ground and by 7.3 m relative to nearby pasture control sites. After the first three years, the rate of decline was near uniform with time. Average salinity of ground water beneath reforestation decreased by 11% (Bari and Schofield 1992).

Engel (1987) reported a series of plantings where the ground water was monitored over 5 years with a series of wells (Fig. 6.6). The trees lowered the watertable immediately beneath but the effect was limited in extent. However, the results indicated how plantations directly affect soil water levels.

Effect of proportion of area planted

A key factor in the rehabilitation of salt-affected agricultural lands is the control of catchment water balance, thus preventing the watertable from rising close to the soil surface.

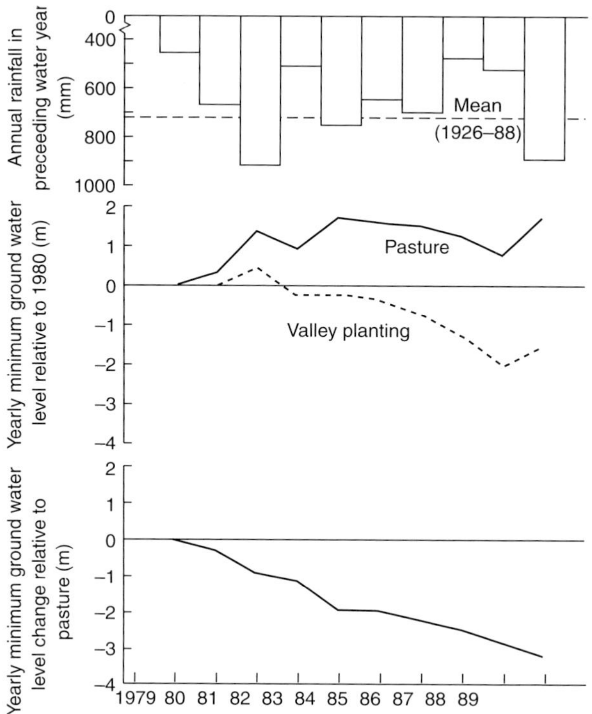

Figure 6.5 Annual changes in rainfall and minimum ground water levels beneath pasture and valley planting (Schofield and Bari 1991).

Where this occurs in addition to soil saturation, evaporation further concentrates the salts. Engineering solutions to prevent emergence of seepages are rarely economically feasible and the most frequently recommended solution to utilise soil water and reduce discharge is to establish trees on up slope re-charge areas (Greenwood *et al.* 1992; Bell *et al.* 1994; Greenwood *et al.* 1994). Upland re-charge areas, however, are generally the most economically important areas and it is difficult to argue for changes in land use from agricultural or grazing activities to extensive tree planting. Direct re-vegetation of catchment discharge zones will also reduce soil water levels, but few workers have attempted this due to the difficulty of the extreme environment. One consideration is that the ultimate success of discharge zone plantations will depend on the success of simultaneous efforts to reduce re-charge in the upper zones of the catchments (Loveday and Bridge 1983). Critically, the problem needs to be addressed on a large scale, such as at the catchment level, with careful planning to locate trees within the landscape to achieve specific objectives.

As discussed earlier, establishment of productive plantations will affect ground water levels beneath the plantation itself, however there are questions as to the effects on adjacent land use and the catchment overall. Specifically, what proportion of the catchment needs to be planted to have a suitable beneficial effect. This is the reverse question to that of clearing effects discussed in Chapter 2 where clearing increased run-off and salinity, and therefore conversely planting would decrease run-off and water salinity. The effect is an interaction between the actual proportion of the area planted, the location of planted areas within the catchment, rainfall, productivity of the plantations, and catchment factors such as hydraulic conductivity

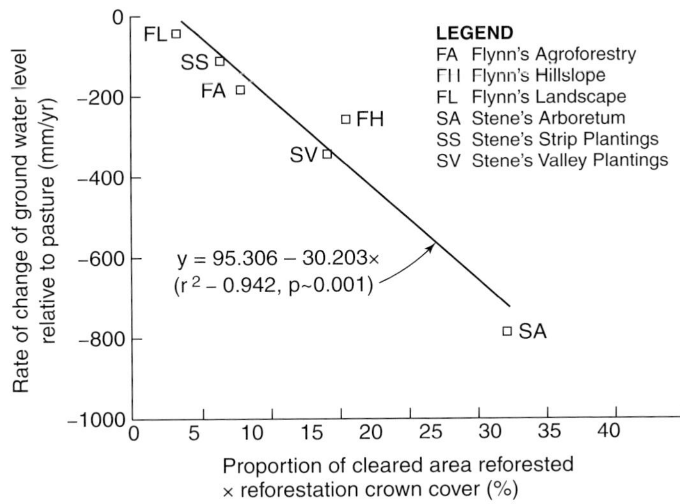

Figure 6.6 Dependence of rate of change of ground water level under reforestation relative to pasture on the product of the proportion of cleared land reforested and crown cover of reforestation (Schofield 1991).

(Loveday and Bridge 1983). Schofield and Scott (1991) provided information from a series of studies on the proportion of cleared areas reforested in relation to the rate of change in ground water level. This was also calculated with a crown cover factor (a surrogate for actual productivity) showing the close relationships (Fig. 6.6). With 25–30% of cleared areas planted, there was a very significant effect (700+ mm rainfall per year) on ground water levels within the catchments. The relationship reported by Schofield (1991) was linear, that is, even small increases in planted area led to a decrease in ground water levels. The later data from older stands indicated a curvilinear effect where it was not until there was a factor of more than 20 that there was a rapid change in watertable depth (Raper 1998).

Plantings need to be sufficiently dense to develop productivity levels which utilise significant quantities of water. Low densities of trees on each hectare may not be productive enough to have a significant effect. Clifton (1992) studied amenity plantings which were developed to control ground water in an area with 435 mm annual rainfall. He considered that 200

Table 6.5 Mean and range of chloride concentrations in grouped wells over a period of 8 years (Greenwood *et al.* 1992).

	Chloride in water (mg/L)	
Wells	**Mean**	**Range**
1, 2, 3	57	24–140
4, 5, 6	87	25–210
7, 8, 9	124	50–510
10, 11, 12	1008	35–9140
X, Y, Z	1560	90–9260

stems/ha was too low and that more than 500 stems/ha were needed. Hence, minimum levels for commercial plantings need to be nearer to the more commonly used 800 stems/ha.

In a multi-species *Eucalyptus* plantation established in 1976, about 12% of a small catchment was planted above a saline seep (Greenwood *et al.* 1992). While the main objective was to rank performance of species in a saline environment, it provided information on changes in ground water salinity (Table 6.5, Fig. 6.4). The changes in ground water salinity were monitored by analysis of samples from piezometers. Larger stores of salt were found in the soil profile down slope than at mid-slope. Chloride in the deep aquifer fell by approximately 20% over 7 years, the decline being greater mid-slope than down slope. The location of the plantation in relation to topography and the aquifers is a key factor in success.

Effects of plantations on soils

The effects of tree plantations on soils have been analysed in a number of studies (for example, Turner and Lambert 1986; 1988) but these have usually been in traditional forest areas. Few studies of effects of plantations are available where soils have been impacted by salt accumulation. Growing trees have been reported to exert ameliorative effects in saline and waterlogged situations by improving physical, chemical and biological properties of the soil. Salt accumulation in the upper layers of the soil is reduced due to greater loss of moisture through transpiration rather than being lost through evaporation from the soil surface. Further, tree shade retards soil evaporation reducing the upward movement of ground water. In areas with high watertables, salt accumulation on the surface is prevented due to lowering of the watertable as a result of increased water use by the trees. Also root penetration increases water permeability within the soil.

The incorporation of organic matter into the soil from root sloughing and litter decomposition leads to the development of favourable physico-chemical properties of the soil. Such changes lead to higher water infiltration and consequently more efficient leaching (Yadav 1980). Such changes are dependent on the particular tree species and their growth rates. For example, some species lead to higher concentrations of calcium in the upper profile as a result of different nutrient cycles.

Typically, salt concentrations in soil profiles can be very high and there are changes over time. Salt content declined from approximately 1 440 000 kg salt/ha (55 000 kg Na/ha) to 60 000 kg salt /ha (23 000 kg Na/ha) in the top 100 cm after 10 years of irrigation (other figures estimated). Considering different species such as *E. tereticornis*, there was up to 27.0 kg Na/ha (including roots) (Ahimana and Maghembe 1987).

The integrated effects of organic matter changes, soil physical changes, variation in nutrient distribution and modification of hydrology as a result of plantation development in saline areas have rarely been studied.

Effects of plantations on water and soil salinity

The effect of establishing plantations affects water and soil salinity simultaneously. The utilisation of soil water will reduce movement of salt to the surface and the ultimate accumulation, however there may be salt accumulation in the root zone where the trees are utilising most of the water. This may be reduced with periodic rainfall events leading to leaching, but

Table 6.6 Change in soil salinity (mS/m) beneath eucalypts over the first two years of growth (Pepper and Craig 1986).

	Soil depth (mm) 0–300	300–600	600–900
September 1976	630 (+ 150)	680 (+ 340)	–
March 1977	1030 (+ 390)	540 (+ 250)	660 (+ 250)
October 1977	1070 (+ 280)	680 (+ 290)	940 (+ 430)
April 1978	1270 (+ 430)	600 (+ 250)	870 (+ 410)
December 1978	2030 (+ 1120)	1020 (+ 410)	920 (+ 220)
2-year change	1400	540	260

this has yet to be determined. Bennett and George (1995) reported significant increases in soil salt over a 9-year period in a *E. globulus* plantation in Western Australia, during which the soil water salinity concentration remained relatively constant. Planted eucalypt species were studied over two years on a saline site where average rainfall was 425 mm (74% falling in the May–September period) and the average watertable depth was 1.2 m. The species grew 1.9 m to 5.1 m over eight years. In the first two years, changes in soil salinity (block average) were as shown in Table 6.6. During this two-year period, the salinity as measured by electrical conductivity had risen by 1400 mm in the soil surface and by 540 mm where the bulk of the roots were located. While the area was relatively dry over the eight-year period, it may be expected that the salt levels rose to more than 4000 mm over eight years leading to substantial long-term stress on the forest crop. The potential changes in salinity will be soil type dependent but will require monitoring with some remedial treatments applied.

An experimental plantation of *E. camaldulensis* and related species was established in a saline valley in 1979 involving saline discharge zones at Whiteheads Creek near Seymour in Victoria. The root zone of this plantation (monitored since 1990) has been shown to receive upward leakage from a semi-confined aquifer at approximately 3 m depth and containing ground water with conductivity from 2 to 3 dS/m. Lowering of the shallow watertable during summer was evident accompanied by a reduction in pressure in the underlying semi-confined aquifer. During spring, soil salinity was relatively stable, increasing uniformly from 2 dS/m near the surface to more than 10 dS/m at 80 cm. As the watertable fell in early summer, the soil salinity fell dramatically at 40 cm and 80 cm, presumably due to leaching by less saline water from the upper profile. From mid-December, salinity increased steadily at 20 cm and below. Superimposed on the general trend were short-term increases in salinity associated with episodes of rain. The size and duration of these fluctuations increased with depth.

Accumulation of salt within the root zone is potentially a major limitation on the use of trees in saline discharge areas. The observations of seasonal variation in salinity at Whiteheads Creek demonstrated that accumulation is subject to the effects of rising and falling watertables as well as leaching by rain. Salinity is seen to vary considerably within the root zone. Simple models of salt accumulation based on annual salt input and annual evapotranspiration are unlikely to produce valid predictions. A long-term accumulation of salinity beneath tree plantations using saline ground water may be expected to occur in the ground water rather than in the root zone. The rate and extent of accumulation will depend on both lateral and vertical ground water flow and salt diffusion in the saturated zone.

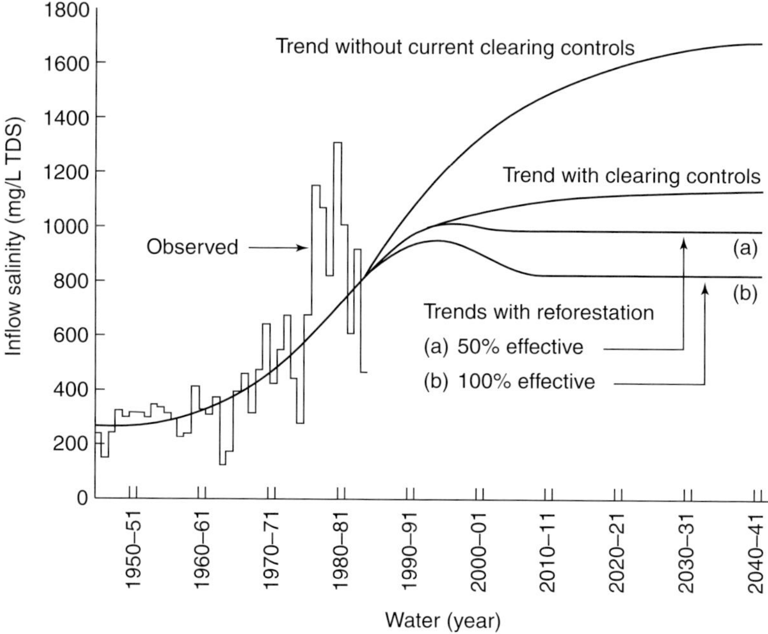

Figure 6.7 Observed and predicted inflow salinities to Wellington Dam for various salinity control measures (Schofield *et al.* 1989). The initial component shows the actual patterns while the potential pattern is indicated together with that for four alternative land management scenarios.

Borg *et al.* (1988) reported on a catchment cleared in the 1970s in which 80% was converted from native forest to farmland. The area was in the 800 mm annual rainfall zone. Between 1977 and 1983, the area was planted to *P. radiata* and *E. globulus*. There was a major decrease in stream run-off and a major change in salt discharge.

Nutrient and salt removals in biomass

Developing plantations take up nutrients and salt from the soil and accumulate them in biomass, with some return to the soil through litter production. Nutrients become immobilised in biomass and when harvesting occurs, a proportion of these nutrients are removed from the site. While the tree biomass appears large, the nutrient content is relatively low and this becomes more so as the stand develops and the tree redistributes essential nutrients from older to newer tissues, such as from developing heartwood to sapwood (Attiwill 1980; Stewart *et al.* 1990; Turner and Lambert 1991). The issue of nutrient removals in forest operations has been widely discussed and in general the conclusions are:

- Nutrients removed in wood from mature stands are small in quantity and will have little effect on site nutrient capital. The reason for this is that a large proportion of wood in mature stands is in the heartwood component which contains small quantities of nutrients.
- Most nutrients contained within the tree are in foliage, branch and bark and their retention on the site at time of harvesting further reduces any impact on nutrient removals.

Conversely, removal of whole trees off site results in the loss of a significant quantity of nutrients.

- Young trees, prior to heartwood formation, have relatively high concentrations of nutrients and there is potential for generally high removals of nutrients.
- Certain tree species accumulate high concentrations of specific nutrients in specific tissues, for example smooth-barked eucalypts accumulate very high concentrations of calcium in the bark, and removal can lead to depletion of this nutrient from the site (Turner and Lambert 1991; Tome and Pereira 1991). This provides a high species effect on site.
- The balance of nutrients in fast grown plantations needs to be monitored, especially for those elements known to be at risk, with ameliorations through fertiliser applications carried out as necessary.

The studies have been carried out in relation to essential nutrients. Salt also accumulates within the plant tissues (see previous sections) and while these quantities may be high from a plant physiological viewpoint, they are low in terms of the total quantities contained within the ecosystem, even if the rooting depth alone is taken into account. The use of trees to reduce the absolute quantities of salt within the landscape through accumulation in biomass and subsequent removal is not a viable option.

Overall effects

The problem of salinity and its control over time are complex issues. A number of similar models are being developed which address aspects of changes in soils and watertables, plant physiological affects and regional changes. These are not outlined here but it is stressed that they have continuing importance for planning and prediction (for example Kaddah and Rhoades 1976; Letey and Knapp 1995; Vertessy *et al.* 1995).

The overall effects related to the area of trees planted, the proportion of a catchment planted, tree productivity, and location of the trees in the landscape especially in relation to aquifers and their depths, are complex but some findings, primarily from southern Australia, may be summarised as:

1 Trees planted in groups can lower watertables, using the level of the adjacent unplanted watertable for comparison.

2 The effect on watertable depth is driven by rate of the productivity of the plantation (productivity per unit area) and will change with forest maturity and consequent change in productivity with age.

3 The quality of the ground water is an important issue and trees planted on fresh to moderately saline watertables can readily lower watertables, the actual change in level depends on rainfall inputs and forest productivity.

4 The most rapid effect will be when more than 25–35% of a cleared area is planted.

5 In general, the effect of trees on the watertable is limited in distance to 40–90 m from the edge of the planted area, this being affected in part by soil hydraulic conductivity.

6 Where trees are specifically planted on discharge–recharge areas, effects can then be over a larger proportion of the catchment. Trees planted in saline or non-saline areas may reduce the recharge and hence affect ground water.

7 The effects of tree plantings on seeps depend on aquifer depth and location of the plantations in relation to the seeps.

8 Solute accumulation can occur above the watertable and is related to the zone of maximum root accumulation. This can change spatially and temporally very rapidly.

9 Trees planted in saline soils may increase the salinity beneath them but the effect varies spatially and temporally.

Concluding comments

The effects of establishing plantations with reasonably high productivity and on a significant proportion of a catchment area are to reduce levels of ground water, stream flow and salt levels. There are also physical and chemical changes within the soil. Such changes are important environmental modifications which need to be taken into account within assessments of commercial plantations. Planting of trees will have an overall positive effect on soils and water as long as a significant proportion of an area is planted with appropriate species. The changes will vary with time and further work is required to suitably predict changes.

CHAPTER 7

EFFLUENT IRRIGATION APPLICATIONS

Issues

Application of effluent to forest plantations is an effective way to utilise waste water and leads to increased productivity. Many effluents are saline and hence need to be appropriately managed. Apart from being a valuable addition to options for management, effluent trials provide a powerful method for assessing salt tolerance of species.

Use of effluent in forest systems

One form of plantation management treatment is to include the use of industrial or municipal effluent. Such treatments differ from normal irrigation programs as they contain pollutants and they are usually targetted at disposal of effluent rather than the management of plantations. In terms of timber production, effluent application systems tend to be relatively small scale, and from a commercial forestry viewpoint, need to be part of a larger timber production program. Only a small fraction of municipal effluent and industrial waste water is recycled in Australia which is surprising considering the country's limited water resources (Allender 1988). In drier areas with moderate amounts of salt in effluent, there is the potential for salt accumulation requiring specific management including the selection of salt tolerant species and irrigation systems allowing periodic flushing of salts below the rooting zone. With suitable management, there can be a decline in soil salinity over a period of time through deeper leaching. Forests, either existing or specifically established for the purpose, are an alternative soil–plant filter (disposal system) with a potential to utilise and renovate large volumes of waste water, while simultaneously producing useful products (Edgar and Stewart 1979). As such, effluent rapidly changes the saline environment as a result of the input of water, salts, and often other nutrients necessary for plant growth. The added nutrients will often lead to high plant growth in the higher saline environment.

One specific form of effluent disposal is where trees have been suggested for irrigation as part of Serial Biological Concentration (SBC) using irrigation water initially on salt sensitive crops and using the more saline run-off to irrigate salt tolerant crops, with the run-off used to irrigate trees finally moving to evaporation pans. To be successful in maintaining sustainable systems, there needs to be leaching where the leaching practices must be greater than or equal to leaching requirements (Tanji and Karejah 1993; Heath and Heuperman 1994).

Effluent application leads to rapidly changing systems requiring very regular monitoring of soil and plant characteristics; and it needs to be recognised that water demand by forest stands will change over time. However, if they are considered as part of an overall commercial timber production system, the management may be modified to fulfill both objectives such as dispose of effluent and produce a commercial crop.

Effluent irrigation and effects on soil salinity

Effluent applications will modify the salinity level and relative ionic composition of soils depending upon the effluent composition and the management schedule for application. For example, ready leaching of a low salinity effluent can lead to additional leaching of salts from the profile and an overall reduction of salt in the rooting zone, while low to moderate applications of a higher salinity effluent can lead to general or specific increases in soil salts. Applications of effluents are significant and need to be considered as a factor in planning, but they are project specific. A major consideration is the specific purpose of the effluent application, whether it is to utilise excess water, or to remove or utilise a specific element or nutrient such as phosphate, or to filter solids out. The site characteristics are also critical requiring reasonable soil water infiltration rates.

Effluent irrigation in woodlots has been undertaken as a method of disposing of both excess water and the contained chemicals. Planning and careful definition of the objectives in relation to the environment are critical to the success of such projects. Components of such planning are the identification of sites suitable for application and the selection of appropriate genotypes. The environment and form of effluent can lead to saline situations or areas with rapidly changing salinity. It is in these situations that much of the work on salinity and tree growth has been conducted and, while representing a relatively small area of plantation, they provide valuable information on performance of species and changes in environmental conditions. There have been a range of studies and demonstration areas in Australia and overseas. A selection of studies demonstrates the value of such trials in relation to salinity.

The elements of effluent systems include:

1 Establishing the specific objectives of such a system (water utilisation, removal of nutrients or suspended material).

2 Determining the environment in which applied (rainfall and evaporation).

3 Estimation of potential water, salt and nutrient loadings.

4 Characterisation of soil physical and chemical properties.

Such systems need to consider the potential for problems through the accumulation of salt, and deleterious effects on factors such as soil structure.

Case studies

Areas where effluent has been applied to forest plantations involve a large number of variables and rather than attempting to incorporate the information into an integrated system, a series of case studies has been analysed. Most of the studies are located where rainfall is comparatively low and evaporation is high for normal plantation forests. Most of the case studies are from Victoria where there has been considerable interest in effluent utilisation.

Case study 1

At Wodonga in the Murray–Darling, a study area was established on trees irrigated with municipal effluent. Rapid growth of trees was demonstrated where water limitations were reduced (Stewart *et al.* 1988; Hopmans *et al.* 1990). The study covered a 4-year period (1980–84) and annual irrigation ranged from 1191–1752 mm. Effective weed control ensured survival. Performance differences between species were demonstrated (Table 7.1) together with assessment of accumulation of nutrients and sodium in both the biomass and the soil. Very significant changes occurred in what was a relatively short period of time. Note that the species with the highest growth of various parameters varied with species, for example the greatest diameter was for Poplar 65/31, the greatest height and basal area for *E. grandis*, and the highest volume for *E. saligna*.

The fastest growing species were *E. grandis* and *E. saligna* with volume mean annual increments of about 32 m^3/ha/year (Table 7.1). Total productivity was estimated using biomass which for *E. grandis* was 95 t/ha and for *E. saligna* 105 t/ha. Accumulation of nutrients in the total biomass differed significantly between species and ranged from 340-400 kg/ha for nitrogen, 4.0–10.4 g/m^3 for phosphorus, 2.1–12.2 g/m^3 for sodium, 22–34 g/m^3 for potassium, 12–61 g/ m^3 for calcium and 4.7–9.3 g/m^3 for magnesium. Two of the slower growing species (*C. cunninghamiana* and *E. camaldulensis)* accumulated more nitrogen, phosphorus, potassium and calcium than *E. grandis* or *E. saligna*, because of their relatively large crown and litter masses. Chemical properties of soils (0–150 cm) were measured in 1980 and again

Table 7.1 Comparative mean growth at four years of age of seven species treated with municipal effluent for 44 months, listed according to decreasing volume (m^3/ha) (Stewart *et al.* 1988).

Species	Stems/ha	Mean dbhob (cm)	Mean dom ht (m)	Basal area (m^2/ha)	Volume (m^3/ha)	Volume MAI (m^3/ha/yr)
E. saligna	1523	12.2	14.3	<u>21.1</u>	<u>134</u>	<u>33.5</u>
E. grandis	1265	12.6	<u>15.0</u>	19.0	126	31.5
Poplar 65/31	620	<u>14.7</u>	14.8	12.4	85	18.9
P. radiata	1161	13.2	6.6	18.7	61	15.3
C. cunninghamiana	1690	8.8	9.5	12.4	43	10.8
E. camadulensis	1690	9.3	8.7	14.3	40	10.0
Poplar 70/51	620	13.1	9.8	9.6	42	9.4

Underlined values are the highest in each column; dbhob: diameter at breast height over bark; MAI: mean annual increment.

in 1984. Irrigation significantly increased soil pH by about 1 unit (the pH range of the effluent being 7–10.6), but reduced salinity throughout the profile. Levels of total phosphorus, exchangeable sodium, calcium and magnesium were increased in the upper profile. Overall, soil chemical properties were not considered to be adversely affected by effluent irrigation.

This study indicated that high growth rates can be achieved as a result of the input of nutrients and water. The location for the species was not within their natural distribution but the effluent was used to ameliorate the environmental conditions. The best growth performance was from species not normally considered to be moderately or highly salt tolerant, so part of the performance was as a result of appropriate management inputs and high growth rates. Nutrient accumulation was not directly a function of productivity, but nutrient accumulation in individual species was a major factor.

Conclusion

High growth rates can be obtained from application of effluent and when managed properly, salt levels can be reduced within the soil profile. The rapid rate of change would mean a sustainable project would require continued monitoring and management adjustments. Where the objective is to remove specific elements from applied water, growth rate alone will not be a sound indicator. Leaching from effluents also results in improvements to aspects of soil chemistry.

Case study 2

Baker (1996) reported on a study using effluent with low-nutrient, moderate salinity applied to a range of species at Altona (near Melbourne in Australia). The trial was initiated in 1988 on 28 ha and the effluent was derived from the process of resin production. The pH of the effluent ranged from 7.7–9.2 (annual average) and the conductivity from 0.75–2.22 dS/m. The soil is clay derived from volcanics and soil EC averaged 3.1 dS/m (partly due to maritime inputs) with high sodicity. There were variable environmental effects from the application with some areas showing general increases in soil EC.

Species planted were *E. grandis, E. globulus, E. botryoides* and *C. cunninghamiana*. Growth was relatively low for all species, for example at four years of age, height of *C. cunninghamiana* was 3.80 m and for *E. grandis* was 6.23 m which can be compared with growth in Table 7.1.

Table 7.2 Growth of a range of effluent-treated eucalypts and other hardwoods in Victoria at 4 years of age (Edgar and Stewart 1979). Sites varied significantly in soil texture.

Species	Mean dom ht (m)			Basal area (m^2/ha)		
Site/soil texture	1	2	3	1	2	3
E. globulus	<u>10.80</u>	<u>9.57</u>	7.60	<u>30.4</u>	<u>16.4</u>	6.3
E. grandis	10.67	7.80	<u>9.50</u>	16.4	9.4	<u>25.1</u>
E. saligna	10.47	7.23	7.03	28.9	12.7	6.9
E. viminalis	9.96	-	5.50	20.4	-	3.4
E. crenulata	9.83	5.57	6.63	20.3	4.8	9.1
E. botryoides	9.53	7.33	8.30	15.8	10.3	22.0
E. cypellocarpa	8.85	7.90	4.83	11.5	12.6	3.4
E. cladocalyx	8.67	6.33	5.86	6.2	7.4	3.7
Poplar I488	8.47	8.43	5.87	12.5	8.7	7.4
Corymbia maculata	7.98	6.30	7.60	8.5	3.5	5.0
Populus deltoides	7.53	5.27	6.37	12.9	3.8	6.4
E. leucoxylon	7.00	6.03	6.20	11.4	3.6	8.0
E. camaldulensis	6.13	5.17	6.50	16.0	6.1	17.3
E. astringens	4.57	5.77	4.77	3.1	5.0	3.7
E. citriodora	1.87	4.63	6.50	16.0	6.1	17.3

Underlined values are the highest in each column.

Nutrient analyses of foliage indicated long-term declines in nitrogen, phosphorus, potassium, sodium and chloride and the trial primarily demonstrated the need for nutritional monitoring and amendment (see Table 9.3).

Conclusion

Effluent with highly imbalanced chemistry can result in nutritional problems and reduced growth rates. Such sites require regular and continued monitoring leading to appropriate amelioration (fertilisers).

Case study 3

The effect of waste water on growth of tree species was assessed in Victoria (Edgar and Stewart 1979). There were interactions between genotype (species) and site (including effluent input) but some species performed well across a range of sites in Victoria (Table 7.2). Mean height provided a general indication of performance but basal area was more critical from a commercial performance point of view as it was more highly correlated with volume or biomass production. *E. globulus* had the highest basal area (30 m^2/ha) on the three sites but there was considerable between-site variability raising the issue of matching genotype to site. No single index of growth can be used on its own for comprehensive assessment and there is a need for continued monitoring of growth.

Stewart *et al.* (1981) reported that nutrient deficiencies or imbalances can result in problems which occur very rapidly. Severe chlorosis was observed after 4 months in *E. globulus* subsp. *globulus*, *E. camaldulensis*, *E. grandis* and *E. botryoides* on one of the sites at Robinvale, located on calcareous soil. The chlorosis was diagnosed as iron deficiency and corrected. At this site the sewage farm is located on calcareous soil.

Considering other trials (Edgar and Stewart 1979, Stewart and Flinn 1984) using both effluent and normally irrigated trees at three sites, there was a general pattern with site differences in that *E. grandis* performed generally well across all sites while some species performed well on specific sites.

Conclusion
High productivity can be achieved across a range of sites with some species performing generally well across a range of sites and others performing well on specific sites. Effluent modifies sites and some species performing well are not native to the area (for example *E. grandis* into Victoria).

Case study 4

Four trials in the Shepparton Irrigation Area provided a range of growth and site information. These are part of the Trees for Profit Program as described by Bren *et al.* (1993); Baker *et al.* (1994); Stackpole *et al.* (1995); and Hamlet and Morris (1996). Details of the sites are:

Cobram has heavy clays with a fine sandy clay loam A horizon. The watertable is greater than 3 metres. Irrigation is with ground water of 0.5 dS/m. Planting was in March 1994 with irrigation shortly after.

Nathalia has a sandy loam with very shallow A horizon over a heavy clay. There are significant sand deposits. Watertable depth is 2–3 metres. The plantation is irrigated from a shallow aquifer with salinity of about 2 dS/m. The trial was planted in August 1993 and irrigation applied in late 1994.

Tatura has heavy clays and clay loams. The profile has a narrow band of sand at about 5 m. Watertable is at less than 2 m. Irrigation is with ground water with a salinity of about 5 dS/m. The trial was planted in September 1993 and irrigation commenced two months later.

Timmering has a heavy clay loam with the watertable at less than 1–2 m. A layer of sand occurs at 4-8 metres. Irrigation is with both channel water (0.2 dS/m) and ground water (10 dS/m). Trees were planted in July 1993 and irrigated in March 1994.

One provenance of each of four species was assessed namely:

- *E. camaldulensis* from Barmah State Forest
- *E. grandis* from Coffs Harbour Seed Orchard
- *E. globulus* from Otway Ranges
- *E. saligna* from Batemans Bay.

Based on mean diameter (Table 7.3), *E. globulus* performed most consistently across all sites. Height did not appear as sensitive an indicator of performance as diameter.

Relationships between increased salinity levels and declines in growth were reported by the authors. While *E. globulus* performed best across all sites, it was found to be the species most sensitively affected by salinity. *E. camaldulensis* was the most tolerant species but showed no consistent relationship with soil salinity. *E. grandis* was quite tolerant of soil salinity levels. The salinity levels predicted to reduce growth by fifty percent at each site are quite low,

Table 7.3 Growth of four species in the Shepparton Irrigation Area. Soil electrical conductivity was determined on a 1:5 extract (Hamlet and Morris 1996).

	Species			
Site	***E. camaldulensis***	***E. globulus***	***E. grandis***	***E. saligna***
Cobram (24 months)				
Mean height (m)	4.15	<u>6.05</u>	5.59	4.74
Mean diameter (cm)	5.23	<u>6.54</u>	5.89	5.22
EC_e range (mS/m)	5.1–11.2	5.5–10.9	6.5–9.7	5.7–9.3
Nathalia (18 months)				
Mean height (m)	2.95	3.37	<u>3.52</u>	3.02
Mean diameter (cm)	3.39	<u>3.87</u>	3.53	2.52
EC_e range (mS/m)	7.0–14.7	7.5–16.5	7.9–15.2	7.9–14.8
Tatura (18 months)				
Mean height (m)	4.91	<u>5.36</u>	4.62	3.99
Mean diameter (cm)	4.00	<u>4.87</u>	4.11	3.68
EC_e range (mS/m)	16.7–22.6	16.6–24.2	16.7–26.7	16.7–27.6
Timmering (18 months)				
Mean height (m)	2.63	<u>3.98</u>	3.58	2.96
Mean diameter (cm)	2.35	<u>3.84</u>	3.18	2.37
EC_e range (mS/m)	9.9–45.6	12.4–47.9	9.1–38.5	12.2–39.9

Underlined values are the highest in the row.

emphasising the effectiveness of salinity in reducing growth rates at concentrations well below those inducing conspicuous toxicity symptoms and tree mortality (Morris 1984). However, the studies demonstrated the value of irrigation projects for controlled studies on salinity.

Conclusion

The study indicated the effect of increasing soil salinity on reducing growth. There is a high degree of between site variation and there are problems of consistent field testing for salinity. The species with assumed highest salinity tolerance, *E. camaldulensis*, did not perform the best.

Case study 5

Werribee Farm, near Melbourne, utilises a range of systems for treatment with municipal effluent. In 1990, 16 ha of trees were established as a demonstration. Four *Eucalyptus* species were planted using different planting spacings to address different end products (for example, short rotation coppice for pulp and much wider spacing for sawlog production after 20 years). Two contrasting soils were planted (Benyon *et al.* 1991). Variation in growth occurred at an early stage between species, provenances, and site (Table 7.4). No recent reports on Werribee are available, however the field designs which include different species, provenances, soils, and planting density provide a strong basis for genetic and site specific management system selection.

Table 7.4 Mean height (m) of four eucalypt species 12 months after planting on two sites irrigated with sewage effluent at Werribee Farm.

Species	Provenance	Initial density (stems/ha)		
		1333	2500	4444
Alluvial soil				
E. grandis	Buladelah, NSW	2.0	2.3	2.2
	Coffs Harbour, NSW	2.0	2.3	2.2
	Gympie, Qld	1.9	2.2	2.2
E. saligna	Yadboro, Qld	2.2	2.1	2.6
	Buladelah, NSW	1.9	2.0	2.5
	Tenterfield, NSW	1.3	1.6	1.6
Basalt soil				
E. grandis	Buladelah, NSW	1.8	–	1.8
	Coffs Harbour, NSW	1.7	–	1.8
	Gympie, Qld	1.6	–	1.7
E. globulus	Flinders Island, SA	1.7	1.5	2.1
	Moogara, Tas.	1.5	1.5	2.0
	Jeeralang North, Vic.	1.6	1.4	2.0
E. camaldulensis	Lake Albacutya, SA	–	1.7	–
	Petford, SA	–	1.2	–
	Isdell River, SA	–	1.1	–

Conclusion

The study demonstrated variation in growth between species and provenance, and hence the need to establish studies in relation to desired end product as affected by spacing. There was an interaction between soil and genotypes. The value of complex trials testing a range of silvicultural and biological alternatives is indicated.

Case study 6

A plantation was established adjacent to a sewage treatment works for the municipality of Forest Hill near Wagga Wagga, New South Wales (rainfall averages 570 mm/yr). Slightly saline, secondary-treated municipal effluent (EC 0.5–1.5 dS/m) alone or mixed with bore water has been applied at three rates for five years (Myers *et al.* 1999). Lack of flushing caused an increase in soil salinity leading to growth reduction (Benyon *et al.* 1996) and this resulted in monitoring and assessment of salt loadings. The variability in results appears very high. Tree productivity changes were only reported in relative terms making comparisons with other studies difficult.

Growth of trees at age 2 years after treatment with effluent at Wagga Wagga was reported by Polglase *et al.* (1994). The EC of the water was 530 EC or 740 EC. At two years, the best twelve species in the trial are listed in Table 7.5, the best two of these being *E. dunnii* and *E. grandis*. Rank was based on survival, stem form, tree height, total leaf area, foliage density, degree of insect and disease damage, physical damage, and tree uniformity. It is

Table 7.5 The twelve best species at age two years in an effluent irrigation trial at Wagga Wagga (Polglase *et al.* 1994).

Species	Provenance	Height (m)	Score	Rank
E. dunnii	Moleton, NSW	6.6	1.0	1
E. grandis	Woondum, Qld	6.4	1.2	2
E. grandis	Coffs Harbour, NSW	6.0	1.3	3
E. botryoides	Bodalla, NSW	6.3	1.5	4
E. regnans	Florentine, Tas.	6.2	1.7	5
A. mearnsii	Bungendore, NSW	6.4	1.8	6
E. badjensis	Brown Mt, NSW	6.7	2.0	7
A. melanoxylon	Burnie, Tas.	4.6	2.0	8
E. saligna	Moleton, NSW	5.8	2.2	9
E. nitens	Toorongo, Vic.	5.2	2.5	10
E. globulus	Huonville, Tas.	6.2	3.0	11
E. bicostata	Mt. Cole, Vic.	6.1	3.5	12

interesting to note that *E. globulus* is a relatively good performer based on growth but ranked lower due to poorer form and uniformity. In such trials, the salinity levels in the soil tend to increase but can be reduced with additional flushing. However such flushing can lead to increased leaching into ground water and while nitrate concentrations below the rooting zone were low, a total of 40 kg/ha/yr of total nitrogen had been lost (Polglase *et al.* 1994).

Studies after only four years indicated there were increases in salinity (for example, from 1000 EC prior to treatment to 2000–3000 EC after 4 years) and these increases were consistent with the salt loadings in the irrigation water. Exchangeable sodium percentage (ESP) increased in all treatments over the four years to levels which are considered to be sodic and causing soil physical problems (Falkiner and Smith 1997). At the time of analysis, no equilibrium had been reached. The accumulation of chloride and boron was considered to have no detrimental effects on growth, although detrimental effects were reported by Benyon *et al.* (1996). The site has been used for modelling of nutrient and water usage (Myers *et al.* 1996).

Conclusion

Application of effluent will lead to high growth rates, particularly of introduced species. There is a need for constant monitoring of soil and water since salt levels change rapidly and affect growth.

Case study 7

At Bolivar Sewage Treatment Works in South Australia, application of secondary treated effluent was commenced to a trial forest plantation in 1991 (termed Hardwood Irrigated Afforestation Trial, HIAT) (Boardman 1991; Hanna *et al.* 1992; Schrale *et al.* 1993; Boardman *et al.* 1996a, 1996b; Shaw *et al.* 1996a, 1996b). Five potentially suitable species were reported from the trial (*E. camaldulensis*, *E. globulus*, *E. occidentalis*, *E. grandis* and *C. glauca)* from 60 species in total. Cost of establishment for drip irrigation was expensive ($12 200/ha). An average of 890 mm effluent was applied to the site which was a sandy loam with a significant airborne salt load on saline subsoil, an annual rainfall of 440 mm

Table 7.6 Biomass (tonne/ha) at five years at the Bolivar effluent study site.

Species	Freshwater	Effluent
E. globulus	108.0	124.9
E. grandis	78.9	75.4
E. camaldulensis	71.0	77.4
E. occidentalis		30.3
Casuarina glauca		120.1

and an evaporation of 2000 mm. The applied effluent was 2000 to 3000 dS/m average EC. Tree species varied in their resistance to insect attack and there were large differences in growth rates. Effluent was compared with freshwater but it was difficult to determine the salt effect as the effluent contained nutrients such as nitrogen and phosphorus. The application involved 272 kg N/ha, 67 kg P/ha, 401 kg Ca/ha, 312 kg Mg/ha, 306 kg Na/ha and 336 kg K/ha. The highest biomass accumulation was 125 t/ha at 5 years for *E. globulus* treated with effluent. This was 17 t/ha higher than the freshwater treatment alone (Table 7.6). Compared with freshwater, effluent increased growth in some species; however this was not universal.

Conclusion

The study showed the differences between species and some of the physiological processes involved. Direct comparisons of effluent and freshwater are valuable and demonstrated the need for more detailed experimental designs over longer periods.

Concluding comments

The studies indicated that effluent trials are often undertaken in relatively low rainfall areas, but where other environmental conditions are suitable for tree growth. The application of effluent adds water but with potentially highly variable ranges of chemical elements. Under such conditions the potential for tree growth is high, and while often in an apparently saline environment, tree species with rapid growth and low salt tolerance often perform well.

Effluent projects using forest plantations have two preferably valuable uses. The first is utilisation of water with chemical elements while providing a valuable commercial product. The systems need to be well planned and managed with selection of suitable genetic material. The second value is to establish a range of environmental conditions, including salinity gradients, for assessment of future genetic material and management systems.

Irrigation with effluent from a range of sources has the potential to increase plantation productivity while utilising what can otherwise be a significant environmental problem. There is the interaction between water, salt and, possibly, added nutrients which require further management and constant monitoring. In systems where effluent disposal is the primary objective, the highest loadings are considered, whereas if timber production is considered important, the optimum may be lower effluent loadings over a larger area. It allows for assessment of tree performance under more controlled conditions and indicates the potential for controlled testing in the future. Monitoring of growth, nutrition and soils in such conditions is critical but the returns and environmental benefits can be very high.

CHAPTER 8

PLANTATION MANAGEMENT IN SALINE ENVIRONMENTS

Issues

The successful management of commercial plantations in conditions of salinity and waterlogging involves optimising conditions for early growth and subsequent management of water and nutrients. The variability of sites together with specific causes of potential site degradation need to be understood to allow for appropriate ameliorative treatments. The benefits from such plantation systems are potentially very high.

Commercial plantations

Successful forest plantation establishment involves suitable analysis of the site and interpretation of requirements. Commercial forest plantations in saline areas are established using a variety of operational systems in which various scales of operation are involved, ranging from traditional large area plantings dedicated to timber production, through to smaller and more diverse agroforestry systems. Regardless of individual size of plantations within a defined region, there needs to be a sufficient total area of productive plantation to support a marketing and products processing system. Issues universal to all plantations relate to availability of planting land, establishment and management of the plantation, harvesting, marketing, and environmental impacts. In saline environments, particularly in drier areas, there are additional considerations of potential access to irrigation water and the potential for environmental rehabilitation of sites.

Plantations systems

The plantation system involves the scale and form of operational management undertaken for a forest area. Within an area, a variety of plantation systems may exist, and although differing in scale, they can be managed as an integrated system if they are producing compatible products. Large scale forest plantations are the traditional form of industrial planting systems. As such, a large area of land needs to be dedicated to timber production and to servicing a specific form of processing, such as a single plant which may be a paper mill or an integrated system involving sawn, reconstituted products, and pulp. A major factor in such developments is the acquisition of large land areas which have suitable productivity levels. Such large plantings in saline areas require careful regional planning since impacts will be significant.

Smaller blocks of plantation within the landscape are an alternative and may be established using a variety of management systems. Further alternatives are agroforestry systems which incorporate commercial tree production within an agricultural system (MacLaren 1995). Often this is linked within a larger venture and the landowner is paid a rental, an annuity or a share of the crop. Such approaches provide income to the landowner while the factor of land cost to the project developer is minimised. In more saline or degraded situations, the landowner also has the additional benefits of environmental and productivity improvements to the property, such as changes in hydrological conditions and improvements in soil properties. In saline areas, such plantings will be planned and undertaken at a farm level where most benefits will accrue, although in aggregate there will be regional implications.

The landowner needs to assess all the benefits to be derived from agroforestry systems including income streams, on-farm financial values, taxation benefits and environmental effects. Financially, the benefits are considered as a balance between production from trees and loss (or gain) in production from alternative existing land uses. If the benefit is positive or zero, the system will be market driven. If the benefit is negative, the system will be driven by the balance between environmental conservation and the landowner's economic considerations. Under such systems of environmental conservation, the tree crop need not give returns that are fully competitive with conventional crops. It is the aggregate costs and benefits that determine whether the tree crop will be competitive. The strategic placement of trees to optimise benefits is essential, and merely planting the trees on areas of land that can't be used for much else is probably of doubtful benefit. Two typical tree layouts are likely to be used in agroforestry:

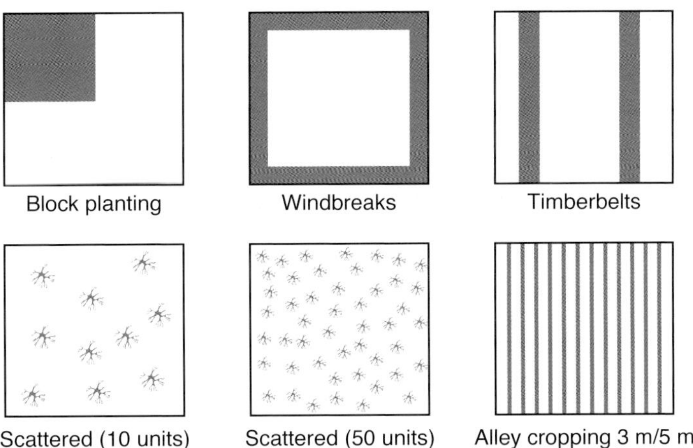

Figure 8.1 Alternative ways of planting 25% of a property to commercial tree crops. The length of the tree crop interface on one hectare of land increases from 200 m for the plantation to 2500 m for the narrow alleys. The water use of trees in plantations is better understood than from scattered trees or strips where advection, lateral root growth and subsurface lateral flow are important (Young 1987).

- Concentrated plantings at strategic landscape positions where particular advantage can be gained, for example valley bottoms, sand plain seeps or break of slope. The position in the landscape for such plantings is a major issue in overall design.
- Dispersed plantings such as belt patterns are suited to utilisation of surplus water across recharge areas, providing extensive shelter or compatibility with cropping. A similar area of planting (ha) can have a number of configurations which individually will have different effects. Some configurations are presented in Figure 8.1.

Plantation establishment and management

Successful plantation project development requires setting product objectives, selecting suitable genotypes and sites, and most critically, selecting the most appropriate management procedures to meet those objectives. Such requirements are common to development of all commercial plantations, but within areas of elevated salinity there are further constraints which need to be addressed, particularly the environmental processes previously outlined, which affect growth. Essentially saline environments are suboptimal for tree growth and there will be changes over time. As such, all aspects of management need to be carefully considered and effects monitored. Within an area for which plantations are to be established there are several considerations as outlined below.

Selection and characterisation of the site

Site selection needs to be undertaken using an analysis of environmental characteristics in relation to growth of specifically identified plantation species. The site selection system needs to meet the requirements of the species. This approach has been carried out for a number of species in traditional plantation situations such as pines, and particularly for short rotation *Eucalyptus* plantations (Turvey 1987; Buckley 1988; Turner *et al.* 1990;

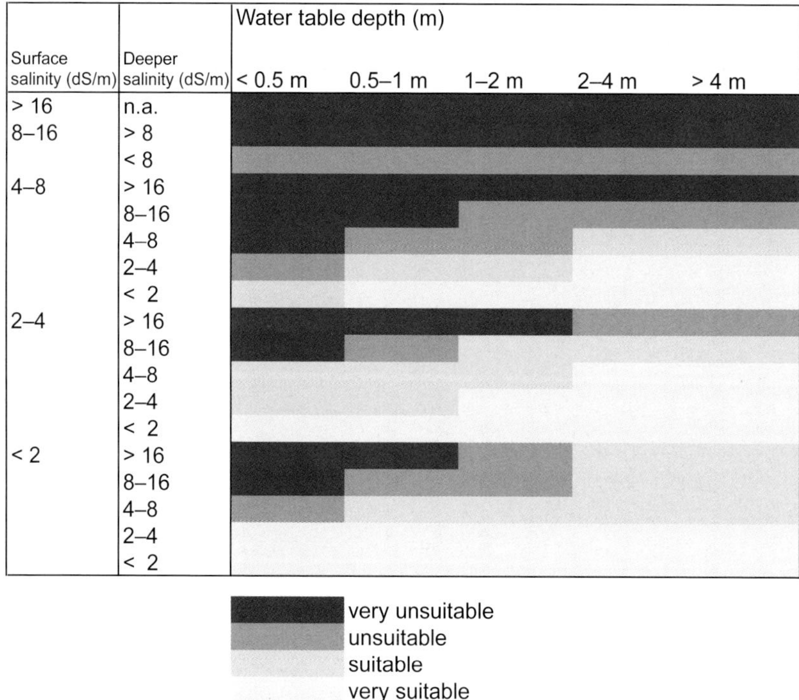

Figure 8.2 Matrix analyses for sites based on Box 1.3. The sites with darker shadings are unsuitable for tree plantations without very significant inputs.

Inions 1991; Edwards and Harper 1996). For example, *E. globulus* performed poorly in Western Australia on sandy soils and soils with gleyed (waterlogged) subsoils. Water storage appeared to be the critical limiting factor for the selection and based on this, selection criteria were developed. A suitable system for use in saline areas was outlined in Chapter 1. The system primarily considers water, salt levels and root restriction and when analysed a matrix for interpretation can be developed as in Figure 8.2 and may be modified for local criteria.

At a more localised scale, location of trees within the landscape where ground water is a consideration requires specific characterisation of the ground water pattern. The position of trees in the landscape will determine access to water and affect quality of water. Trees in up slope areas usually have limited access to ground water while contact with ground water can be intermittent at midslope and replenished by lateral movement. Permanent high ground water table, often with poor quality, occurs toward the bottom of catchments and will have impacts on growth. As plantations develop, there will be a change in soil water patterns and the previously saturated bottom slopes may become very suitable for plantation growth. The selection of sites hence requires analyses of present conditions and some interpretation of potential changes.

Preparation of the soil

Soil or site preparation may include removal of existing vegetation together with the residues, ploughing, ripping, mounding, and levelling in cases where flood irrigation is employed. The preparation will be determined according to existing soil conditions and planned future operations. The primary objective is to modify soil physical conditions so that they are optimal for continued root development of the growing crop. As pre-treatment conditions vary with each soil and site, site preparation will need to be matched to the site. In the case of saline and waterlogged sites, mounding may have to be considered to establish initial growth, or in other areas modification of the site may be required to accommodate irrigation systems.

Land preparation techniques and cultural operations that have shown significant improvements in tree performance, particularly for plant survival, include:

- Mechanical ripping of soil with tynes to break up 'plough hardpans', clay layers or calcium carbonate hardpans, or manual digging or mechanical augering of planting holes. The soil replaced in the hole may be chemically amended with gypsum and fertilisers (Grewal and Abrol 1986; Gill and Abrol 1991).
- Use of saucer pits and furrows as 'water harvesting' techniques. In this method, seedlings are planted in a depression designed to collect rainfall. The technique is appropriate in drought-prone and/or saline soils where a high watertable does not exist.
- Use of mounds in areas affected by waterlogging to give seedlings access to a volume of aerated soil above the watertable. The 'trench-ridge' technique is widely used in Pakistan and India and the 'double-ridge' mound has been effective in aiding shrub and tree growth, though not necessarily overall survival. In the double-ridge mound method, the trough between the ridges where seedlings are planted collects rain which leaches salt from the soil below the trough, that is the seedling root zone. These mounds can be manually or mechanically constructed (Malcolm and Allen 1981; Yadav 1989; Hoy *et al.* 1994; Marcar *et al.* 1995).
- Provision of subsurface drainage. This can be provided to remove excess water either by 'vertical drainage' such as pumping from tube wells, or 'horizontal drainage' such as deep open ditches (trenches), slotted pipe (tile or plastic) drains and mole drains, or both. In extensive forestry, this is probably prohibitively expensive but it may be important in localised situations.
- Application of mulches to reduce evaporation (drying of the surface soils) and accumulation of salt at and near the soil surface. Mulches include straw, cotton burrs, vermiculite, peat, sand and polythene.
- Addition of organic and inorganic fertilisers. Sodic soils are often very low in organic matter, nitrogen, zinc and other nutrients. Gypsum can be mixed with sodic soil to replace excess sodium with calcium, but this will only be successful if sufficient water is provided to leach excess sodium beyond the root zone. Improved soil structure and drainage are the outcomes of such treatments. Significant survival and growth advantages accrue from use of farmyard manure in addition to gypsum, either mixed with original soil in auger holes, or broadcast onto the soil, and lead to more efficient use of fertilisers by reducing soil pH (Grewal and Abrol 1986; Singh 1989).

The depth of soil treatments and the conditions under which they are applied are critical. In saline sites where mounding is used, there are effects to consider due to the different heights of mounds (Fig. 8.3). Such preparations will be site and region specific and considerable testing is required. Specific modifications to mounds have been found to be successful in certain circumstances. For example, Fitzpatrick (1994) reported on successful double-ridged mounding of saline sites in Western Australia (Fig. 8.4). He stressed that the treatments needed to be carefully evaluated as the costs in creating big mounds may be prohibitive. Successful results were considered to be due in part to the seedling trough collecting

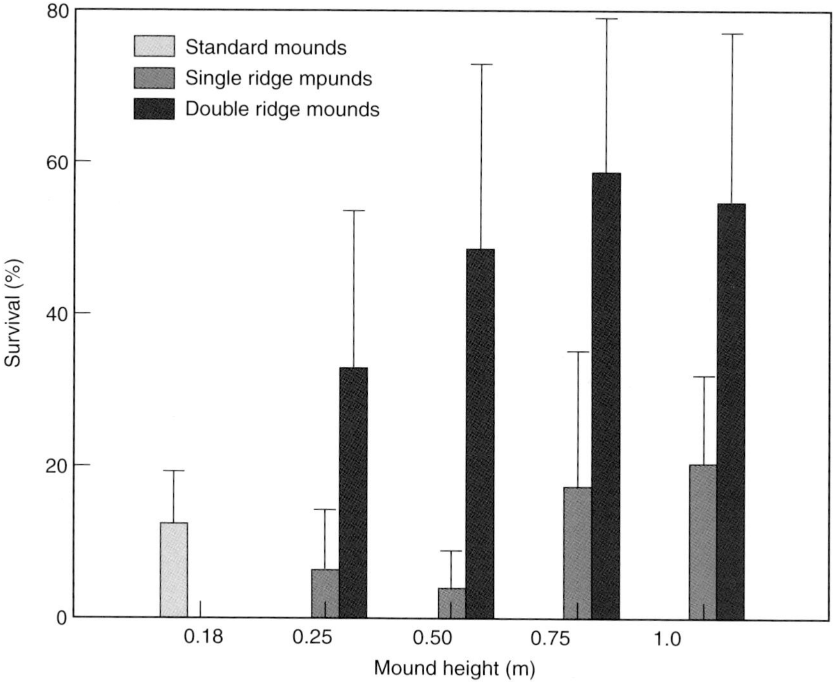

Figure 8.3 Survival of seedlings three years after planting in single and double ridge mounds. Data are averaged for two species (*E. camaldulensis* and *E. largiflorens*) (Pettit and Froend 1992; Fitzpatrick 1994).

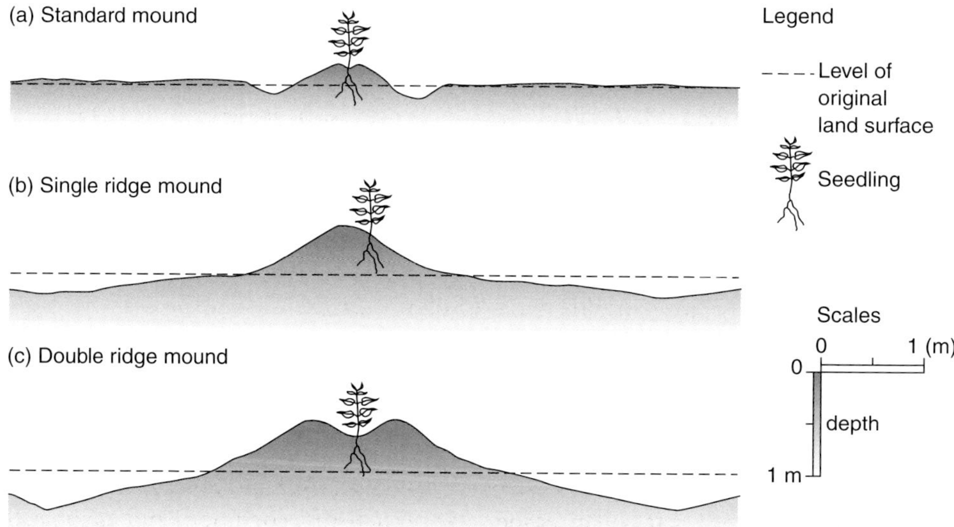

Figure 8.4 Alternative mounding on saline sites (Fitzpatrick 1994).

rainwater which facilitated leaching of salt, and to the trough directing the occasional summer rainfall to the seedling root zone.

Plant competition management systems

Management of vegetation competition specifically through control of weeds, is an important objective in optimising tree growth. Weeds compete for light, water and nutrients. The type and density of weeds will be key factors in determining the control method, that is whether they are woody weeds, grasses or other forms of vegetation. Most of the control measures will be in the early stage of plantation development, since with rapid tree growth and crown development it can be expected that weeds will be suppressed.

As planted trees grow they compete for water, nutrients and in some cases light. At the time of establishment, weed species need to be identified and controlled. In some circumstances, cover crops or other inter-plantings may be used and hence the objectives need to be established for managing vegetation competing, or potentially competing, with the plantation crop. Management may include physical or chemical pre-planting or post-planting treatments and will be specific to the particular location.

In situations where irrigation is involved or the watertable is close to the surface, the use of chemical control measures needs to be carefully planned and implemented to minimise ground water contamination from such chemicals.

Where irrigation is being undertaken, the potential environmental effects of chemical weed control (for example, ground water contamination) need specific careful evaluation.

Planting stock selection

Selection of genotypes for planting is one of the most important for success of the project and is determined by the desired end product and expected growth rates from the site. In addition to genotype, the type of planting stock and source of material will also be decided. Quality control of planting stock assists with early rapid growth which is an important factor for subsequent development. Alternatives include the use of seedling or vegetatively produced material and the level of improvement of the seed source, including 'wild' seed selection, open pollinated or control pollinated material. The cost of the genetic material increases according to the level of improvement, however benefits in growth rates, uniformity and quality can be very significant. In traditional plantation systems, improvement is based on growth rate, tree form and related phenotypic characters. The characteristics will also be the basis for selection in saline areas, although the pressures causing growth changes will be different. These will be important in saline sites. The quality of the planting stock should be specified, detailing height, root development, foliage form, health, and nutritional status.

The decisions on genetic material and planting stock will directly affect both establishment and long-term success, so the decisions need careful consideration.

Timing of planting

The season when trees are planted has direct effects on survival and growth of plantations. The reasons for selecting a particular season relate to moisture conditions and minimisation of stress by heat or dryness. In many temperate areas, planting is undertaken in winter or early spring when plants are not actively growing and moisture stress is minimised. In

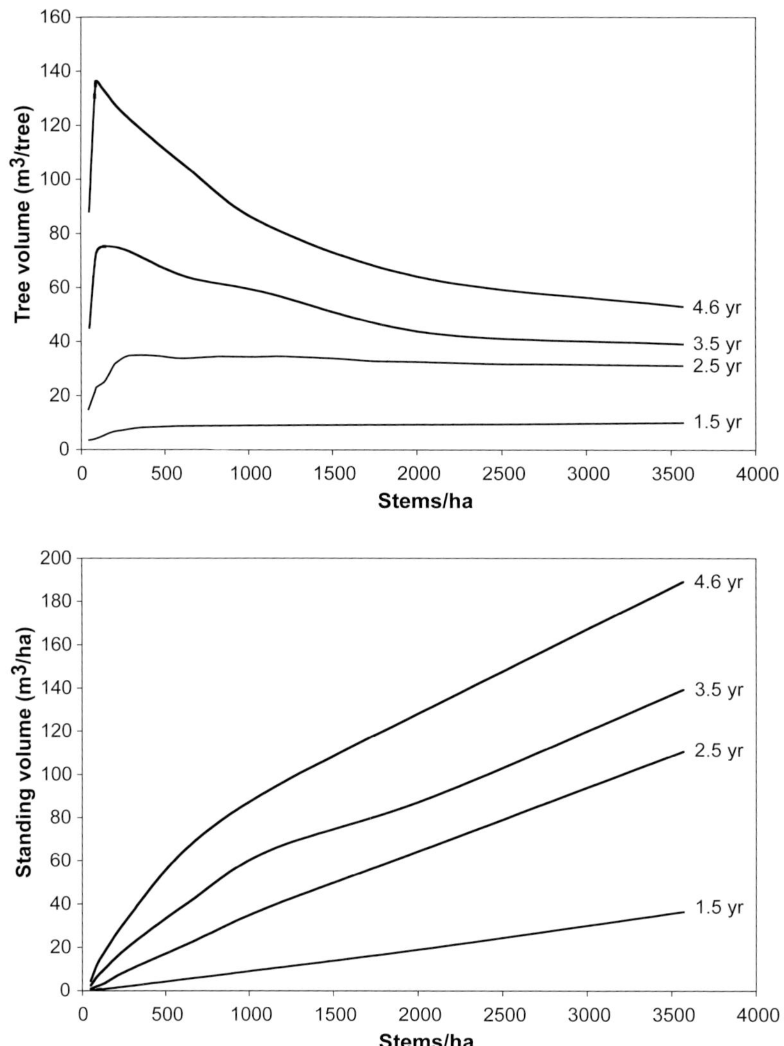

Figure 8.5 Individual tree volume (m^3/tree) and total accumulated volume (m^3/ha) for an *E. grandis* stand at different spacing over four different ages (Ryan 1993a). Individual tree sizes rapidly decline with more than 1000 stems per hectare. Reductions in numbers of stems per hectare will reduce costs for plant purchase, fertiliser application, pruning and other factors, but need to be balanced against productivity, competitive effects and quality.

warmer areas, planting is undertaken from winter through to late summer depending on seasonality of rainfall.

Planting area configuration and initial plant spacing

The number of stems planted per hectare directly affects both the early net productivity and the individual tree size. A large number of stems per hectare will produce high initial total productivity but the individual stems will be relatively small. Such a regime is often used for short rotation pulp crops. Smaller numbers of stems per hectare will provide lower overall productivity, but larger individual tree sizes are often used with high quality genetic stock where there is a target high quality product (sawlog or veneer log). The more densely planted sites will, over time, have high mortality (self thinning) so that net productivity will be reduced but the individual trees will increase in size. Insufficient stems per hectare will lead to under-utilisation of the site, weed development and other problems. A typical effect of spacing is shown in Figure 8.5 for different aged plantations of *E. grandis*. Where planted material is of doubtful quality, higher planting densities may be used to allow early thinning to remove poor quality individuals and retain the better quality stems to grow on further.

The configuration of tree planting (the actual spacing of trees), will be a key part of the overall management system. In short rotation pulp crops it may not be an issue but if access for subsequent treatments is required, the spacing will need to be wide enough to allow for equipment access. Spacing and configuration, such as spacing of mounds, need to be carefully planned as soil preparation systems will be directly affected.

The planting area configuration (blocks, strips, contour planting, etc) will be determined by site characteristics and overall management objectives including the expected environmental impacts. Within such areas, the spacing of individual trees and the configuration of that spacing will determine the end product. On an area with several distinctly different soils, several planting regimes may be utilised. Close spacings will be used where the product is pulp, while wider spacing with a configuration enabling access for machinery will be used where high quality products, grown on longer rotations, are the objective.

Irrigation

In many areas, tree growth is limited by water availability. Irrigation is not normally used in broadscale plantations partly due to costs but also because of long-term water availability. Typically, irrigation has been used on smaller projects for effluent disposal. Irrigation may be used in forest plantations at the establishment stage in areas with rising watertables to allow root development to the watertable. Irrigation may be used for a limited period for establishment purposes.

In drier saline areas or where the project is partly for effluent re-cycling, the irrigation system needs to be planned at time of site evaluation and the schedule implemented at time of planting. The options of flood, channel, spray or drip systems will be dependent on the area and objectives of the project together with water restrictions and availability, and relative costs. The irrigation management systems will include the need for adequate leaching to reduce salt accumulation in the root zone. Soil characteristics evaluated for routine irrigation need to be considered, including hydraulic conductivity or infiltration rates.

The quality of the water will be a significant factor in the subsequent health of the crop and change of salt in the soil, and hence evaluation of water quality will be important. General guidelines for use of irrigation water have been developed in a number of areas and under Australian conditions; there are guidelines available (ANZECC 1992). The Australian ratings for salinity are related to potential impacts on soils and the sensitivity of plants to a range of chemical elements. The species reported did not include forest tree species, but selected characteristics of water quality have been included as a basis for monitoring. The following details are from ANZECC (1992).

Table 8.1 Chloride concentrations in irrigation water causing foliar damage.

Sensitivity	Chloride (mg/L)	Affected Crop
Sensitive	< 178	almond, apricot, plum
Moderately sensitive	178–355	grape, pepper, potato, tomato
Moderately tolerant	355–710	alfalfa, barley, corn, cucumber
Tolerant	> 710	cauliflower, cotton, safflower, sesame, sorghum, sugar-beet, sunflower

Major ions in irrigation water

Bicarbonate

No quantitative guideline is recommended for bicarbonate because the potential hazard of bicarbonates is influenced by other soil and water characteristics. The bicarbonate hazard is high in water with low-salinity applied to sandy and silty loam soils.

Chloride

Chloride is essential for the growth of plants. However in excess it can be toxic, depending on the sensitivity of the crop and the irrigation method chosen. Irrigation water containing more than 100 mg Cl/L should not be used for sensitive crops. The maximum chloride concentration should be set according to the sensitivity of the crop (see Table 8.1). In general, most woody plant species (stone fruit, citrus, avocados) are sensitive to low concentrations of chloride, whereas most vegetables, grain, forage and fibre crops are less sensitive. There are two ways of inducing chloride damage. First, the chloride ion can be taken up by the roots and moved upwards to accumulate in the leaves. Excessive accumulation may cause burning of leaf tips or margins, bronzing and premature yellowing of leaves. Secondly, direct foliage absorption of chloride from sprinkler irrigation can cause damage, especially in fruit trees which are the most sensitive.

Sodium

The major problem associated with sodium in irrigation water is its tendency to adversely affect soil structure. The magnitude of this effect can be related to the relative proportion of sodium to calcium and magnesium ions in the irrigation water (SAR or Sodium Absorption Ratio). Sodium damage due to applied irrigation water can occur in two ways: either through alteration of the soil structure or via direct toxicity to the crops. The direct effects of sodium concentrations in irrigation water (expressed as SAR) on different plants are shown in Table 8.2.

Excessive sodium in irrigation water relative to calcium and magnesium can adversely affect soil structure and reduce the rate at which water moves into and through the soil, as well as reduce soil aeration. The relation of sodium to calcium and magnesium is expressed as:

$$\mathrm{SAR} = \frac{[\mathrm{Na}^+]}{\dfrac{\sqrt{[\mathrm{Ca}^{2+}] + [\mathrm{Mg}^{2+}]}}{2}}$$

where $[Na^+]$, $[Ca^{2+}]$ $[Mg^{2+}]$ are concentrations in milli-equivalents per litre. If calcium is the predominantly adsorbed cation, the soil tends to have a granular structure which is easily worked and readily permeable. However, when adsorbed sodium exceeds 10–15% of the

Table 8.2 Tolerance of crops to sodium.

Tolerance	SAR of irrigation water	Crop	Condition
Very sensitive	2–8	deciduous fruits, nuts, citrus, avocado	leaf tip burn, leaf scorch
Sensitive	8–18	beans	stunted, soil structure favourable
Moderately tolerant	18–46	clover, oats, tall fescue, rice	stunted due to nutrition and soil structure
Tolerant	46–102	wheat, lucerne, barley, tomatoes, beets, tall wheat grass, crested grass, fairway grass	stunted due to poor soil structure

Table 8.3 General guidelines for salinity of irrigation water (ANZECC 1992).

Water class	Comment	Electrical conductivity (μS/cm)	TDS* (mg/L)
1	Low-salinity water can be used with most crops on most soils and with all methods of water application with little likelihood that a salinity problem will develop. Some leaching is required, but this occurs under normal irrigation practices except in soils with extremely low permeability.	0–280	0–175
2	Medium-salinity water can be used if moderate leaching occurs. Plants with medium salt tolerance can be grown, usually without special measures for salinity control. Sprinkler irrigation with the more saline waters in this group may cause leaf scorch on salt-sensitive crops, especially at high temperatures in the daytime and with low application rates.	280–800	175–500
3	High-salinity water cannot be used on soils with restricted drainage. Even with adequate drainage, special management for salinity control may be required, and the salt tolerance of the plants to be irrigated must be considered.	800–2300	500–1500
4	Very high-salinity water is not suitable for irrigation water under ordinary conditions. For use, soils must be permeable, drainage adequate, water must be applied in excess to provide considerable leaching, and salt-tolerant crops should be selected.	2300–5500	1500–3500
5	Extremely high-salinity water may be used only on permeable, well-drained soils under good management, especially in relation to leaching and for salt-tolerant crops, or for occasional emergency use.	> 5500	3500

* *Total Dissolved Solids (mg/L) = 0.68* × electrical conductivity (μS/cm)

total exchange capacity of the soil, the clay becomes dispersed and puddled when wet, lowering permeability and forming a hard impermeable crust when dry. Most researchers agree that the problems of soil permeability increase when SAR approaches 10.

Total dissolved solids (salinity)

The salinity or concentration of Total Dissolved Solids (TDS) in irrigation water is an extremely important water quality consideration. An increase in salinity causes an increase in the osmotic pressure of the soil solution, resulting in reduced availability of water for plant consumption and possible retardation of plant growth. Recommended guidelines are shown in Table 8.3.

With adequate drainage, salt accumulation in the soil can be controlled to an extent by the rate of application of water. If the sum of applied irrigation water and rainfall is lower than evaporation and plant consumption, an accumulation of salt in the main root zone will result. Appropriate irrigation management will allow application of sufficient excess water (leaching fraction) to move a portion of the salts out of the root zone, while not causing excessive increases in the ground water table. Relative tolerance of species has been reported in terms of sensitivity to saline irrigation water (Table 8.4).

Boron

Boron in relatively small amounts is essential for the normal growth of all plants; however this element can be toxic when present in excess. Crop species vary both in their requirement and in their tolerance of excess boron. The concentration of total boron in irrigation water should not exceed 0.5 mg/L for sensitive plants. Boron sorption plays an important role in determining soil solution concentrations. Concentrations of 1–2 mg/L usually occur in the soil solution only when the adsorptive capacity of the soil is saturated. Sensitive crops may therefore be grown in natural to alkaline soil for a longer period without harm from irrigation water containing 2 mg boron /L.

pH

Irrigation waters with extreme pH values may cause indirect problems. Values less than 4.8 can cause solubilisation of aluminium, manganese or heavy metals in concentrations large enough to be toxic to plants if applied to acidic soils over a long period. Waters having high pH values (greater than 8.3) may contain high concentrations of sodium, carbonate and bicarbonate. These ions may effect growth and soil conditions, as discussed previously.

Fertiliser applications

Fertilisers are applied to plantations to overcome obvious nutrient deficiencies and also to stimulate growth even when there is no apparent deficiency. Fertiliser applications can result in substantial growth increases when applied at various stages in a rotation. The magnitude and duration of the response will depend on stand condition and on the form, rate, and method of fertiliser application. There are four possible reasons for applying fertilisers to plantations (Knott and Turner 1990; Turner *et al.* 1996b) and these directly determine the type and amount of fertiliser to be applied and the subsequent type of response:

1. *To assist in successful plantation establishment.* Usually this is achieved by applications of phosphatic fertilisers to individual trees, or nitrogen and phosphorus (NP), or NP plus trace elements at planting. The treatment is applied in relatively small amounts of fertiliser, adjacent to an individual tree at a time when root development has been

Table 8.4 Relative tolerance of ornamental trees to saline irrigation water (from ANZECC 1992).

Water class	Electrical conductivity (µS/cm)	TDS (mg/L)	Tree species
1–2	0–800	0–500	*Camelia* spp., *Magnolia* spp.
3	800–2300	500–1500	*Podocarpus (elatus)*
4	2300–5500	1500–3500	*Acacia longifolia, Callitris cuppressiformis, E. kondinensis, E. longicornis, E. loxophleba, E. occidentalis, Ficus* spp., *Hibiscus* spp.
5	> 5500	> 3500	*Casuarinas* spp., *E. sargentii Tamarix* spp., *Atriplex* spp.

restricted. Typically, 100–250 g of fertiliser per tree will be applied in forms such as superphosphate or diammonium phosphate.

2. *To correct recognised nutrient deficiencies* (or imbalances) which result in economically significant levels of tree mortality, deformity, or losses in growth potential. In these cases, treatments are generally applied where deficiency symptoms are obvious or determined by foliar analysis. In Australia, phosphate, boron and potassium have been the main nutrients falling within this category.
3. *To increase the rate of wood production in actively growing stands* where no nutrient deficiency symptoms are obvious. These stands are usually at pre-commercial age and selected for treatment on the basis of foliage analysis or an understanding of the nutritional status of stands on specific sites. Typical treatments are 500 kg/ha of triple superphosphate broadcast applied to a 5-year-old stand.
4. *To maximise the economic benefit* associated with increased growth immediately after the release of a stand by thinning. Application of appropriate fertiliser leads to rapid crown development and stimulus of growth. Typical treatments will include 200 kg N/ha and 50 kg P/ha broadcast applied immediately after thinning.

Saline sites generate additional problems. These are related to the imbalance of some elements, particularly calcium and magnesium or the reduced availability of some elements due to elevated soil pH (reduced acidity). Such conditions can be modified by the use of apropriate fertilisers. At time of planting, ameliorants such as gypsum or lime may be considered to address soil physical problems in addition to nutrient deficiencies or imbalances.

Fertiliser treatments will increase productivity of forest plantations at different stages of stand development. There is no single treatment for fertiliser application, and the specific applications will need to be matched to the site to overcome nutrient deficiencies and imbalances. The formulation, rate, method of application and timing of application are all factors that will have an impact on both short- and long-term productivity. In terms of agroforestry management, formulations may be modified to accrue long-term benefits from residual fertiliser nutrients derived from previous management.

Many trials and demonstrations using fertilisers have been undertaken and there is a basic understanding of the effect of site characteristics and basic nutritional problems. Trials

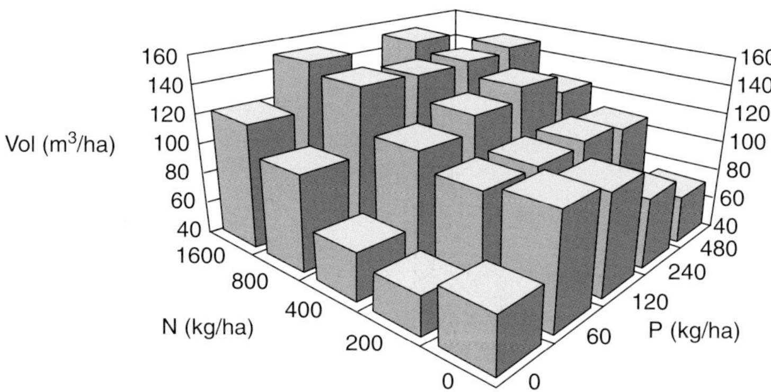

Figure 8.6 Stem volume (m^3/ha) of *E. grandis* (Coffs Harbour seedlot) in response to five levels of N and P fertiliser at age 5.6 years.

provide a wide range of data such as those in Figure 8.6 (from Cromer *et al.* 1995). The optimum fertiliser combination was 60 kg P/ha and 800 kg N/ha and growth was almost double the volume production in the control.

Application of fertiliser will interact with other types of management treatment such as weed control and thinning. For example, fertiliser effects may be limited where control of competition is low and there may be interactions with the type of site preparation (Fig. 8.7). The reason for such interactions will vary in detail depending on the site but in general, fertiliser responses will occur when stand modifications such as reduced competition or other resources are more available.

Pruning

Pruning involves removing branches and with it a proportion of foliage cover within a stand. The primary reason for pruning is to improve quality through production of knot-free wood termed 'clearwood'. Pruned sawlogs are sold for a much higher price than equivalent sized unpruned logs. If a plantation is established for producing pulp where clearwood is not an issue, pruning does not need to be undertaken. Also, it may not be required in production of high quality wood in selected species. Even though pruning is expensive and slows tree diameter growth, there is a trade-off in loss of productivity for improved quality. Pruning also assists in modifying the form of the trees where large branches are involved.

Pruning is carried out in lifts which represent the height to which pruning is carried out from the ground. The pruning considers the diameter of the stem to which pruning is to be carried out and also needs to address the proportion of live crown removed (or left) as this will affect the continued productivity. After pruning, branch stubs occlude (grow over) with wood produced subsequently, but there is a period of growth before occlusion occurs, leaving a 'scar' in the wood. Pruning is routine in species such as *P. radiata* but less common in *Eucalyptus* spp. The timing, height of lift, and quality control need to be addressed for each species, similarly the economics. A concern with some species has been the potential for pest and disease infestation on the branch wounds after pruning. Such a problem needs to be considered in relation to species and location.

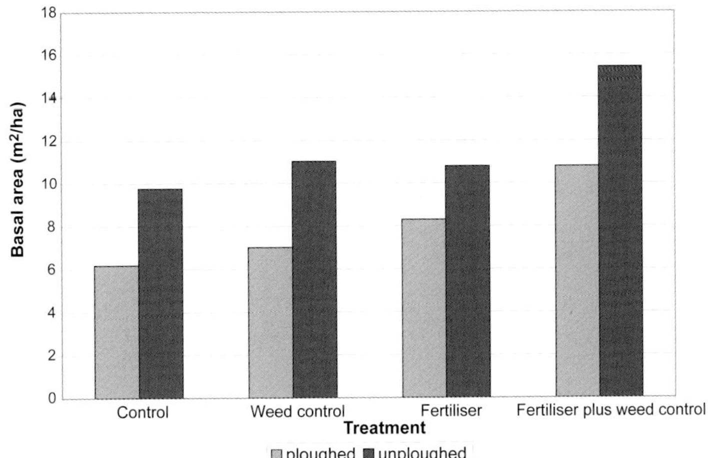

Figure 8.7 Effects of fertiliser, weed control and site preparation on diameter (10 cm from ground) of *E. grandis* at 13 months after planting. The 'control' had no added fertiliser or weed control (Bonny 1991).

Thinning

The thinning of stands and the schedule are important where higher quality products are planned. In short rotation pulp crops, thinning and pruning are not undertaken. A large number of thinning regimes exist and will not vary on saline land from those on more traditional plantations. Extensive thinning will modify, for a period of time, the amount of water usage and hence impact on the hydrological cycle. Increased infiltration and run-off may be expected.

Other land uses

Establishment of plantations on farms will usually be in conjunction with other forms of land use. This may be diverse and management will need to be integrated specifically for the individual property.

Protection

Protection of plantations involves management of factors such as insects and disease, and protection from fire. Insect damage can be expected to be a problem especially in *Eucalyptus* plantations. The types of insects and level of damage will vary with the geographical location, site, species and management system, and require monitoring with expert support. The effects of damage can be expected to be reduced in the longer term through the use of tree breeding but this tends to be specific to different insect groups, and processes and controls need to be understood.

Insect damage varies with genotype and there are significant interactions with management systems. Interactions between species and management were shown by Stone (1993). Nitrogenous fertiliser and insecticides were applied to young *E. grandis* and *E. dunnii* plantations. Fertiliser increased growth and the interactions are shown in Table 8.5. *E. dunnii* tended to have a higher resistance to insect attack than *E. grandis*, the effect for *E. grandis*

Table 8.5 Interactions between fertiliser use, insecticide applications and genotype on impacts of insect damage (Stone 1993). Growth was the mean diameter at breast height (cm) at 26 months. The figure in brackets is the score of foliage damage by insects.

		Species	
		E. dunnii	***E. grandis***
No fertiliser	No insecticide	7.02 (0.2)	5.37 (0.79)
	Insecticide	6.52 (0.19)	6.80 (0.30)
Fertiliser (NH_4NO_3)	No insecticide	9.53 (0.18)	5.62 (0.61)
	Insecticide	9.77 (0.18)	9.96 (0.26)

showing the interaction between fertiliser applications and insecticide. Psyllid damage was primarily limited to *E. grandis*.

Monitoring and assessment

Monitoring and assessment need to be undertaken to determine stand performance, assess any health and/or damage problems, and modify management operations. This aspect is of more critical importance than in conventional plantations and is discussed further in Chapter 9.

Research requirements

Large scale commercial plantations, particularly in new areas and with highly variable saline conditions, will have initial and ongoing research requirements to improve their productivity and value. The type of research will vary depending on the amount of available information and the developmental stage of the plantation, and this will affect the priority of such work. Examples of such areas of work are shown in Table 8.6 assuming a new plantation program developing in a new area for a short rotation pulp, using at this time an unspecified species.

In many areas there is extensive knowledge on genotypes, sites, management, products and their interactions. However, in areas where there are potentially greater stresses on forest plantations through salinity or water deficits, information and understanding is much more limited. Research is required in saline areas at all scales and for both short- and long-term studies.

Concluding comments

Plantation management on saline and waterlogged sites requires a greater level of planning and performance monitoring than on traditional sites. However, establishment of plantations at a landscape level will lead to productive, integrated systems with ongoing environmental benefits. It is critical that research is undertaken and integrated at different scales.

Table 8.6 Types of research and development requirements in relation to commercial forest plantations on saline areas.

Type of research and development	Status	Priority
Site selection and analysis		
Site evaluation (for example, see Box 1.2)	low	very high
Genotype evaluation		
Species assessment (across sites)	low	very high
Provenance testing (across sites)	low	very high
Tree improvement		
Breeding and selection	moderate	high
Evaluation of improved material	moderate	high
Silviculture and stand management (treatments across all sites)		
Site preparation	moderate	high
Competition management	moderate	high
Initial spacing	high	low
Irrigation	high	moderate
Establishment fertilisers	moderate	very high
Later age fertiliser	high	moderate
Pruning	low	low
Thinning	low	low
Protection		
Pests and diseases assessment	moderate	moderate
Pests and diseases management	moderate	low
Monitoring assessment		
Growth	reviewed after establishment	
Nutrient status	reviewed after establishment	
Soil water	reviewed after establishment	
Soil salinity	reviewed after establishment	
Pests and diseases	reviewed after establishment	
Wood properties		
Wood properties	reviewed when stands growing	
Environmental effects		
Soils	low	high
Water	low	high
Other	to be reviewed	

CHAPTER 9

ASSESSMENT AND MONITORING OF PLANTATION HEALTH AND ENVIRONMENTAL VALUES

Issues

The productivity and health of plantations established in saline and waterlogged areas can be either incrementally or rapidly affected by factors in the less than optimum environment. Sound management should involve regular and systematic assessment of a range of parameters and they should be used to modify current and future management procedures. In saline areas, changes to plantation health and productivity may occur more rapidly than in more traditional areas, hence monitoring is more critical.

Plantation monitoring

Establishment and management of plantations requires significant investment in land, planting, and ongoing management and maintenance. Sound management requires monitoring and assessment of a number of key factors to establish that the plantation is performing as expected, and to determine whether there are developing problems which require amelioration. A major objective is to diagnose developing problems at a very early stage when minimal losses have occurred and amelioration is relatively easy. By assessing several key components, explanations of change may also be provided to assist with future management. Key areas to be selectively monitored in plantations, especially those established in saline and waterlogged areas, include:

Establishment performance: Assessment of early growth and survival to determine success of establishment.

Growth performance: Determination of early growth rates of trees.

Competition: Assessment of weed development and types of weed species present.

Health: Assessment of infestations by pests and diseases, their extent and effects on growth and quality.

Nutrition: Monitoring of nutritional and salt status of stands at different ages to determine if nutrient amendments are required and degree of salt accumulation.

Soils: Assessment of salt accumulation in soils and organic matter development.

Soil water: Determination of watertable levels and salt concentrations.

Run-off or streamwater: Assessment of quality and quantity of streamwater run-off.

Such monitoring systems need to be systematic and integrated for the purposes of cost effectiveness and efficiency of reporting. The systems will vary depending on the size of the project but ultimately will determine the level of success of the project. To complement the process of monitoring, a series of standards need to be developed by which changes or measured levels can be evaluated. Such comparisons against baselines will provide the basis for making decisions on required modifications to management. For example, if growth is reduced and weeds are assessed as high, the management decision may be to reduce weed levels.

Additional to plantation monitoring, there is a need for operational monitoring which essentially audits the effectiveness of operational procedures. Such monitoring is undertaken as part of maintaining the quality of plantation management and should address factors, such as whether site preparation, fertiliser applications, weedicide treatments, planting rate, planting stock quality and so on meet pre-determined specifications. This is not detailed further here but recognises routine operational efficiency is critical for project success.

Further details on plantation monitoring are provided below including, in some cases, selected baselines.

Establishment performance

Post-planting evaluation: Establishment performance in saline areas requires inspection immediately after planting in order to evaluate plant condition post-planting.

Baseline: Five percent mortality.

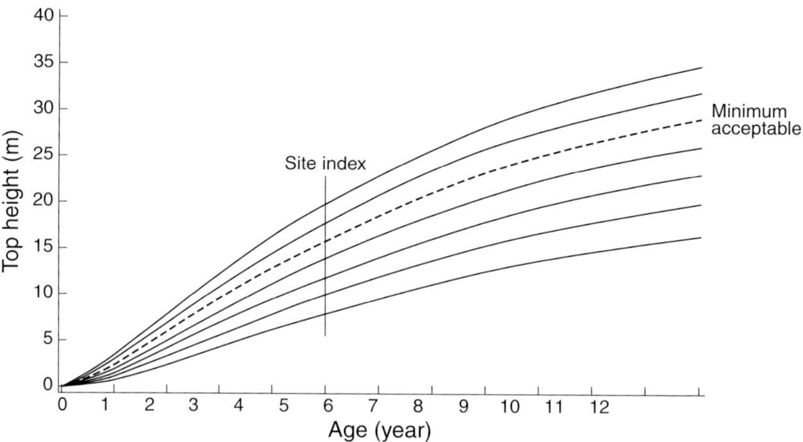

Figure 9.1 Early age height growth curves for *E. globulus* grown in Western Australia (Inions 1991). The dotted line represents the minimum acceptable growth for that area.

Management response: Immediate refill planting up to 6 months after establishment.

Comment: The planting of trees into saline environments may lead to immediate water stress. The evaluation needs to comment on cause of death if extensive in order to modify future procedures. One area may include evaluation of nursery stock quality.

Growth performance

Growth analysis: Quantitative analysis of survival and growth should be undertaken at or near 12 months after planting. This should include a quantitative estimate of survival, height and condition of the plants. Evaluation should be stratified to cover the variation in the plantation (species, soils, topography, etc). Analyses of growth at later age will need to be undertaken as a separate procedure.

Baseline: Five percent mortality is baseline for survival. Projected height, developed from a height age curve, for example of the type shown in Figure 9.1 by Inions (1991) in Western Australia, is required for comparison.

Management response: Evaluation of causes of higher mortality or reduced growth and health. Causes of reduced growth need to be analysed in relation to nutrient status, soil condition and health surveys. Where there is evidence of nutritional problems (symptoms), foliage analyses should be undertaken with the potential for fertiliser amendment, however it should be recognised that nutritional problems usually affect growth rates but not survival.

Competition

Assessments in Year 1 and Year 2: Assess development of weed species in terms of cover, growth, species or type. The type of assessment and quantification will vary with type of species, for example whether they are grasses, flatweeds, woody weeds, etc.

Baseline: The baseline is related to the initial objective of treatment, for example total control of weeds on planted strips in the first 12 months or total control on the whole area.

Management response: Where excessive competition is developing, immediate control treatments are needed.

Health

Forest plantations are susceptible to a wide range of pests and diseases which directly affect the productivity and the quality of the crop. These vary according to tree species and location and may be indigenous or exotic. Plantation management requires regular assessment to determine any damage followed by use of local specialists to determine cause and possible treatment.

The effects of insect pests and diseases have been indicated as a damaging factor in a wide number of plantations and studies (for example in *E. camaldulensis* in Queensland affected by blister sawfly (Fraser *et al.* 1996)).

Periodic assessments: Systematic assessments to determine level and type of pests and diseases. There will also need to be further and immediate responses if there are outbreaks of pests and diseases.

Baseline: Highly variable, but needs to be related to estimated acceptable productivity losses.

Management response: If populations of pests and diseases are near or at unacceptable levels, expert advice is required as to suitable control measures.

Nutrition

Foliage chemical analyses to determine plant nutrient status have traditionally been carried out for three reasons, namely to diagnose existing nutrient problems, to predict possible developing nutritional problems, and to monitor crop nutrient status to maintain optimal productivity. In addition, for saline areas there is the requirement to assess for adverse or potentially toxic elements.

Elemental concentrations in forest plantations vary with site, management, stand age, season, genetics, individual trees within the stand, type of tissue, and position of sampled foliage within the crown. Most programs sample foliage. To interpret nutritional status, especially in relation to site and management, the other variables need to be standardised and calibrated. The value of foliage sampling programs is very high when sampling is standardised and well calibrated. This means specifying the tissue to be sampled, position in the tree, and season. Typically, this is current foliage from upper parts of the crown taken in a slow or non-growing period. Chemical analysis is carried out in laboratories which have appropriate experience with forest materials and where standard material is included for comparison.

Critically in fast grown plantations, there can be rapid changes in nutrition. This is especially true for stressed areas such as with high salinity where irrigation is undertaken or there are other ameliorative undertakings. The rate of change will be also affected by the soil types on which the plantation is growing. Such an example is shown by Kennedy (1992) in Table 9.1 from woodlots in South Australia irrigated with winery effluent with salinity 15–69 mS/cm (pH 6.2–8.3). Leaf chemical analyses were monitored on three blocks where trees were:

- chlorotic on poor soil
- on average soil with no chlorosis and
- on the best soil and with best growth.

Table 9.1 Foliage nutrient concentrations in an effluent-treated *E. camaldulensis* plantation (Kennedy 1992). Figures in **bold** indicate the nutrient is deficient and probably limiting growth and underlined indicate potential nutrient toxicity. This is the form of information received for nutritional monitoring. Corrective NPK solid fertiliser was applied in winter 1991.

	Poor soil				Average soil				Best soil			
Nutrient	**4/1988**	**5/1990**	**6/1991**	**8/1992**	**4/1988**	**5/1990**	**6/1991**	**8/1992**	**4/1988**	**5/1990**	**6/1991**	**8/1992**
N (%)	2.45	1.50	1.80	1.50	1.75	1.80	**1.30**	1.50	2.10	1.70	1.70	1.70
P (%)	0.23	0.16	0.13	0.14	0.15	**0.11**	**0.08**	**0.11**	0.16	**0.10**	**0.10**	0.12
K (%)	1.76	0.74	0.65	0.98	0.91	0.78	**0.35**	0.68	0.98	0.78	0.58	0.72
Ca (%)	0.83	1.50	**0.10**	1.25	0.32	1.10	**0.10**	0.78	0.29	1.00	0.70	0.76
Mg (%)	0.70	0.38	0.27	0.30	0.33	0.28	0.22	0.22	0.29	0.31	0.24	0.28
Na (%)	0.69	0.27	0.35	0.35	0.50	0.27	0.38	0.36	0.41	0.23	0.39	0.35
Cl (%)	0.91	0.27	0.39	0.31	0.50	0.35	0.36	0.31	0.52	0.35	0.38	0.31
Zn (ppm)	19	41	47	29	15	21	21	14	15	18	27	23
Mn (ppm)	526	540	236	229	129	460	319	272	119	667	354	272
Cu (ppm)	–	19	16	11	–	10	9	10	–	7	9	9
Fe (ppm)	–	73	–	–	–	100	–	–	–	71	–	–
B (ppm)	–	245	118	98	–	350	68	184	–	270	93	97

Table 9.2 Sufficiency and toxicity concentrations of foliage nutrients reported for several *Eucalyptus* species (J. Turner and M.J. Lambert unpub. data; Herbert and Schonau 1990).

Nutrient	*E. grandis*		*E. pilulari*	*E. globulus*		*E. maculata*
	Australia	South Africa		Australia	South Africa	
N (%)	2.4	2.9	1.6	1.4	2.0	1.7
P (%)	0.12	0.14	0.08	0.09	0.14	0.10
N/P	20	(21)	20	16	14	17
K (%)	0.60	0.75	0.50	0.70	0.60	1.00
Ca (%)	0.10	0.16	–	–	–	0.29
Mg (%)	–	0.35	–	–	–	0.09
Total S (%)	0.17	0.20	0.12	0.11	–	0.18
Mn (ppm)	–	600	–	–	–	22
Fe (ppm)	100	110	–	–	–	40
Zn (ppm)	–	18	–	–	–	12
Cu (ppm)	–	12	–	–	–	6
B (ppm)	20	32	20	20	–	–
B toxicity (ppm)	200	–	200	–	–	–
Na toxicity (%)	1.00	–	0.80	1.00	–	–
Cl toxicity (%)	0.80	–	0.60	–	–	–

Nitrogen, phosphorus and magnesium declined with time and there was some dilution of nutrients in the rapidly growing stand on the best soil. This indicates the requirement for calibration of foliage nutrient analyses. Critically, the data show the rapid rate at which nutritional changes can take place and hence the requirement for consistent monitoring.

Periodic assessments: Usually foliage samples are taken in the third year, but in saline areas foliage sampling annually for the first five years would be valuable.

Standards or baselines: Standards for foliage nutrients are generally applied to allow interpretation of chemical analyses (see Table 9.2 for example). They are generally fixed levels even though the nutritional requirements vary with factors such as age and productivity class. Standard concentrations are reasonable for a first approximation but in intensively managed plantations there needs to be significant refinements.

Management response: Where major nutrients (N, P, K) are shown to be low, fertiliser applications need to be applied. The levels of such treatments will be dependent on site, species and stand condition. Other nutrients will need more careful assessment as to the cause and may require very specific treatments (for example such as chelated iron).

Soils

The objective of monitoring soil properties is to detect potentially developing problems through changes in absolute levels and by following the trends. Establishment of plantations will lead to soil changes in a number of ways, such as through utilisation and redistribution

Table 9.3 Soil electrical conductivity, tree height and foliage nutrients in a developing *E. grandis* plantation (Baker 1996). The figures in **bold** indicate the nutrient is probably limiting growth while figures underlined indicate toxicity.

	Year					
	1991	**1992**	**1993**	**1994**	**1995**	**1996**
Electrical conductivity (EC_e dS/m)						
Site 8			1.33	1.20	1.37	0.86
Site 3			0.36	1.20	1.13	1.39
Tree height (m)			1.59	3.48	4.69	6.23
Foliage nutrients						
N (%)	1.82	**1.17**	**1.12**	1.37	**0.93**	
P (%)	0.19	0.12	**0.11**	**0.11**	**0.07**	
N/P	9.6	9.8	10.2	12.5	13.3	
Calculated SO_4-S (ppm)		505	439	269	568	
K (%)	1.00	0.75	0.63	0.61	**0.31**	
Na (%)	0.45	0.42	0.35	0.28	0.23	
Cl (%)	0.86	0.80	0.58	0.75	0.31	
Na/Cl (in me%)	0.81	0.82	0.94	0.58	1.10	
B (ppm)	24.9	45.0	50.1	59.3		

of elements and accumulation of organic matter. Where irrigation or effluent is applied, rapid changes in soil conditions and plant growth can be expected. These changes need to be monitored in a variety of ways, and the method needs to be selected according to the area at most risk. For example, where irrigation water is applied to a project, rapid leaching of elements may be considered a key issue and therefore needs to be monitored.

Monitoring of plantation soils generally indicates slow rates of change in most properties. Soil organic matter initially declines after disturbance followed by longer term accumulation, although in many traditional areas soils may be 15–20% lower in organic matter at the end of a rotation than at the beginning (Turner and Lambert 1999). If soil organic matter is a key indicator, changes may need to be measured every five years with possible evaluation of carbon into labile and stable carbon components. Other elements will need to be selected for monitoring on an individual project basis.

Results of soil chemistry for effluent-treated surface soils (Baker 1996) show variability in patterns across the site and in relation to other parameters (Table 9.3). The study indicated that soil parameters need to be related to tree and environmental change as in the relationship between soil conditions and foliage nutrients. The long-term application of effluent leads to reductions in foliage phosphorus and fertiliser applications may restore them to initial levels. Potassium declines as does calcium. On the stressed sites sodium preferentially accumulates, while on the best sites sodium and chloride are in balance. At the time of this report, chemical ameliorants, probably single superphosphate, would be an appropriate treatment. The chloride at these sites fluctuates. One-off analyses can demonstrate extremes but in marginal environments the development of trends and patterns through consistent

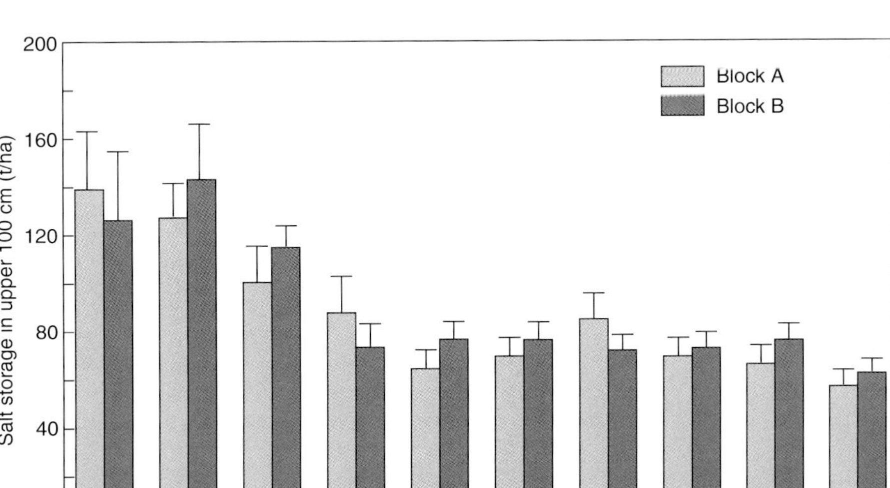

Figure 9.2 Salt storage over time in the upper 100 cm of the soil profile in a plantation irrigated with saline ground water. Two separate blocks of trees were monitored (Morris *et al.* 1994). Each column is the mean of 16 to 18 observations. Bars represent one standard error.

assessment is of greater value. While there is evidence of some increase in soil salts with time, this is not reflected in the foliage of the growing trees. Patterns of salt storage can be assessed over time, for example see Figure 9.2 (Morris *et al.* 1994).

Baseline: The baseline will usually be the result of the initial sampling and patterns will be developed in relation to this. The initial results may indicate a problem and subsequent analyses will then assess effectiveness of the treatment.

Soil water

Watertable depth

Soil water beneath plantations needs to be monitored routinely for a number of factors. The depth to watertable at a number of key sites should be defined. This will fluctuate seasonally and annually but the longer term trend is an important factor.

Baseline: The depth to watertable at time of planting.

Soil water quality

Soil water quality needs to be monitored for salt changes in saline areas. Accumulation of salt indicates long-term stresses will occur. Further, where ground water is quite high, application of chemicals such as weedicides or fertilisers may need to be assessed to determine configuration. Where effluent is applied, there is movement of contaminants in the profile, such as losses of nitrate.

Baseline: The baseline will be the concentration level at the start.

Assessment of soil water needs to be considered where chemicals (fertilisers, weedicides, etc) are being used either regularly or at high levels. There should be both regular and periodic sampling and this should be coordinated with the overall assessment of environmental change. Types of monitoring patterns are shown in Figure 9.3 from Schofield and Scott (1991).

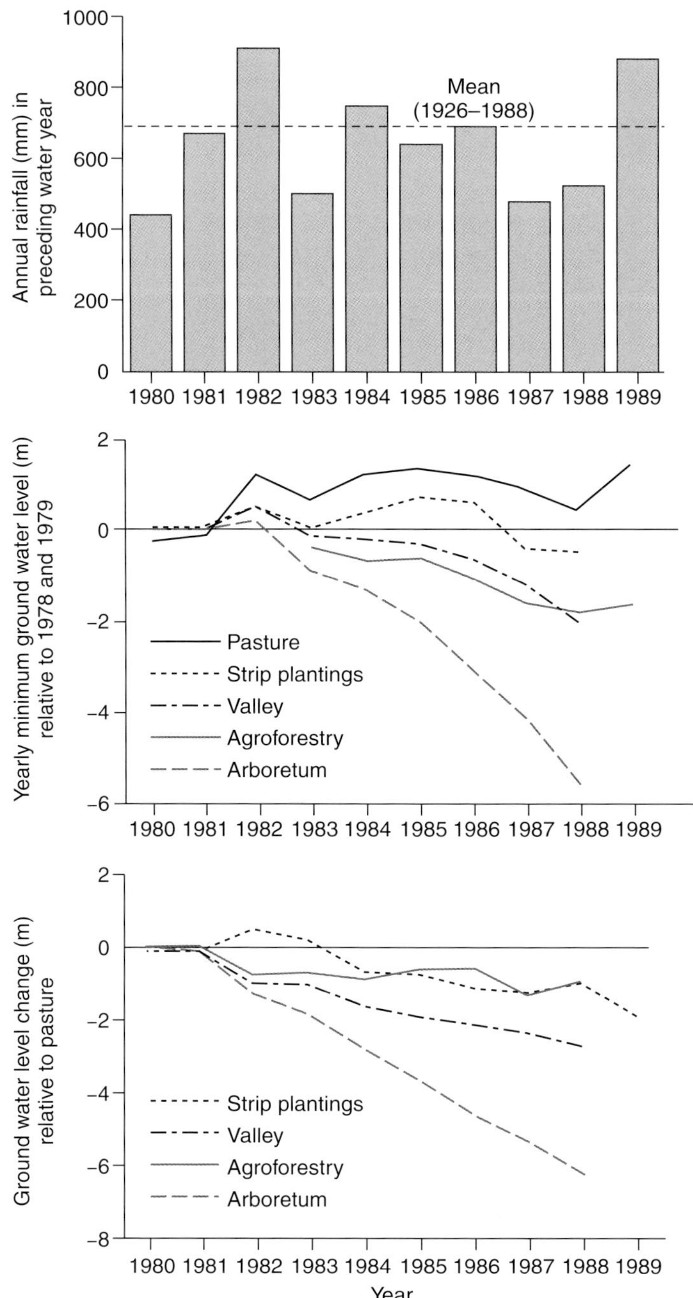

Figure 9.3 Changes in ground water level relative to the ground surface for various reforestation strategies (Schofield and Scott 1991).

Run-off or streamwater

This requires strategies for run-off water assessments which will be determined by the scale of the operation. They may include assessments of streams from surrounding land use and need to consider both spatial and temporal variation (Turner *et al.* 1996c, 1996d). An overall monitoring program of plot trials for water quality, soils, and tree nutrition has been outlined by Duncan *et al.* (1997).

Run-off water quantity and quality are important issues for consideration as they affect downstream users and there are potential effects on the aquatic environment. Any assessment of run-off water quantity will need regular and long-term monitoring to account for the high level of variability. Changes in water quality may occur both in terms of salinity and nutrients, and plantings may need to be located to optimise improvement in water quality.

Baseline: Establishment time quality and quantity.

Assessment of products in plantations

Establishment of commercial plantations requires that accurate assessment and prediction of products is made. Comparisons need to be made between planned levels of productivity and the monitored levels of inputs and outputs of the plantation to ascertain profitability. The main product will be timber although the actual products (pulp, sawnwood, pole, veneer) may be tracked individually. There is also the potential for trading of carbon accumulation.

The measurement of wood and carbon can be undertaken in the same plots and may be part of the same estimation. A number of plots need to be established, sufficient to provide a precise unbiased estimate of productivity. The methods for plot establishment are set out in a number of manuals and are not repeated here. Information from the plots can be used in stand volume functions and in biomass prediction functions. The biomass can be converted to carbon by the use of a factor. Soil carbon also needs to be monitored as it will change with differing land management practices.

Baseline: The predicted values and rates of carbon accumulation according to site and species.

Concluding comments

Plantations require significant investment but commercial and environmental benefits are very large. The more intensive the management the greater the risks, but these can be reduced by monitoring, evaluation and management response.

Systems for monitoring and assessment need to be planned as part of the overall management system and integrated within it. The greater the stress on the trees and/or the higher the growth rates, the greater the requirement for monitoring. The value of improved management will offset costs of monitoring and produce effective plantation systems. This monitoring will be in addition to routine operational monitoring (for example, quality control) and product assessment (for example, wood, carbon).

CHAPTER 10

CARBON ACCUMULATION IN FOREST PLANTATIONS

Issues

As forest plantations grow, carbon accumulates in both the woody tissues and in the soil. In addition to direct on-site environmental benefits of carbon accumulation, there are indirect benefits through their effect on the 'greenhouse gas', carbon dioxide. Interest is developing in this area for the potential to sell the quantities of carbon accumulated, and this would be a source of direct and immediate income for landowners while further stimulating plantations establishment.

Introduction

Establishment of forest plantations is considered one of a range of options available to provide a sink for the greenhouse gas, carbon dioxide (CO_2). The temperature of the Earth's surface is dependent on the balance between energy coming in from the sun and that outgoing from the Earth. Increasing carbon dioxide in the atmosphere leans the balance towards warmth, as some of the outgoing infra-red energy is absorbed by carbon dioxide in the air and is re-radiated back to the ground. This is similar to the effect of glass in a greenhouse, hence the term. A natural greenhouse effect keeps the present average surface temperature of the Earth at 15°C, which is some tens of degrees warmer than would be expected if there were no greenhouse gases in the atmosphere (Steering Committee of Climate Change Study 1995). Carbon dioxide emissions have been increasing since the late 19th century with the commencement of industrialisation as industry started to use fossil fuels such as gas, oil, and coal as a source of energy. Two changes occurred, one being the large increase in energy usage and the associated increase in emissions, and the other is the change from renewable fuels, especially wood, to fossil fuels. The effect of these emissions together with others such as those derived from changes in land use has led to increasing concentrations of a range of greenhouse gases including carbon dioxide.

The impact of increased atmospheric concentrations of carbon dioxide has been widely predicted to result in increased temperatures, modified and more erratic weather patterns, and increases in sea levels through the melting of polar ice-caps. Indications are that changes will differ geographically and general predictions of change are difficult. The extent and degree to which these will occur are not fully understood. Calculations of future emissions of carbon dioxide based on a range of scenarios suggest a concentration of double pre-industrial levels will occur by 2100 and that temperature will rise by an average of 2°C (range 1.5–4.5°C) and sea levels by 50 cm by that time. It is estimated that the rise in temperature would be smaller at the equator and greater at the poles, and increase with an increase in global average rainfall (Steering Committee of Climate Change Study 1995; Fries 1997).

Atmospheric carbon dioxide is currently increasing at about 0.4 percent per annum, and now constitutes approximately 355 parts per million by volume (ppmv) compared with 280 ppmv in pre-industrial times (Pearman 1989; Watson *et al.* 1992). Estimates of the global biomass pool range from 550 to 830 billion tonnes of carbon (bt C) with the other major pools being the atmosphere (720 bt C), oceans (38 000 bt C) and in fossil fuel reserves (6000 bt C) (Adger and Brown 1994) although there is considerable variability involved in these estimates. Forests are very significant in the accumulation of carbon and are considered to contain 80% of the above-ground terrestrial biospheric carbon and 40% of terrestrial below-ground carbon (Kirschbaum *et al.* 1996). Annual emissions from deforestation are estimated to be 1.6 bt C (1 t C = 3.67 t CO_2) and an estimate of fossil fuel emission is 5.4 bt C.

Globally, there is a common concern about elevated carbon dioxide levels and hence ways are being sought by which the cause and impacts may be reduced (for example, United Nations Framework Convention on Climate Change 1992; Kyoto Protocol to the Framework Convention on Climate Change 1998). The most direct way to reduce emissions is by reductions in the use of fossil fuels. In conjunction with this, there is the more efficient use of fossil fuels through modified transport and use of more efficient industrial processes. Rather than using fossil fuels, alternative energy sources could be used and they may include wind, solar, hydro, wave, nuclear and biomass. More efficient industrial uses include materials that require less energy in

Table 10.1 Carbon emissions resulting from manufacture of common building materials (after Buchanan and Honey 1994).

	Carbon released		Net carbon (kg/m^3)	
Building material	**(kg/tonne)**	**(kg/m^3)**	**Stored**	**Emitted**
Treated timber	44	22	250	–228
Glue laminated timber	164	82	250	–168
Structural steel	1070	8132	15	8117
Reinforced concrete	76	182	0	182
Aluminium	2530	6325	0	6325

their production. For example, in the building industry there is a wide range in the energy used in production of materials. High energy materials include aluminium, steel and concrete. To produce one tonne of steel, approximately two tonnes of carbon dioxide are released (Table 10.1). Timber requires a much lower energy input for manufacture and many of its uses (such as in buildings) represent a direct storage of carbon. Timber may be viewed as an environmentally compatible building material in terms of production and storage of carbon, and is a logical end point in conjunction with plantation forests.

Potential carbon trading

The Kyoto Protocol approved the principle of trading in carbon within specified carbon emitting or accumulating activities. Within the Kyoto Protocol (Article 3.3), the activities specified for meeting national commitments in relation to carbon accumulation are direct human-induced land use change and various forestry activities which are limited to afforestation, reforestation and deforestation. Such activities are only considered relevant when undertaken since 1 January 1990. Afforestation (in the Glossary of the Revised 1996 IPCC Guidelines for National Greenhouse Gas Inventories) is considered as the establishment of forest where forest has not historically existed and reforestation is defined as establishment of forest where forest has historically existed, but where the land has been converted to another use. Both of these definitions are relevant to the establishment of plantations in areas with elevated salinity. Deforestation has not been defined but would probably involve the conversion of a forested area to an alternative land use so it is a total loss of forests (that is, not left to regenerate).

Under the Protocol, parties may participate in emissions trading. Hence, the losses and sequestration of carbon need to be understood and quantified in a precise and consistent manner. Plantation forests are an identified sink and appropriate for carbon trading although at this time the systems of trading, requirements and actual values are still not fully understood. Carbon trading schemes potentially have the following elements:

- Systems to quantify the amount of carbon actually or potentially sequestered.
- Certification to issue carbon certificates and provide assurance to purchasers and regulators that the due processes have been complied with.
- Access to a group of carbon emitters who need to secure carbon credits.

- Access to a group of forest growers who can deliver carbon sequestration rights.
- A trading system which allows carbon certificates to be traded.

As trading will be international and the quantity of carbon to be sequestered is so large, it is likely that a significant component of forest sequestered will be in the Southern Hemisphere.

Carbon accumulation within plantations

Forests, both plantations and regrowth, are being considered for both increased sequestration of carbon and for providing an alternative fuel source (Barson and Gifford 1989). Industries which emit large quantities of carbon dioxide, such as coal burning power stations, will have to reduce their carbon dioxide emissions or they will be penalised in some way. A 'carbon credits' trading system for emissions will allow tradable permits to be bought and sold among different industries and companies. High carbon dioxide emitting industries will be able to buy carbon credits from more efficient industries including carbon storage industries such as plantation forestry, in order to reduce the quantity and cost of emissions.

Carbon dioxide has an atomic weight of 44 and there is 27% carbon in one molecule of carbon dioxide. In plant sugars, the amount of carbon is typically closer to 43%. That is, if the quantity of organic matter contained within a forest can be estimated, then the quantity of carbon may be estimated using the conversion factor. In many cases, the assumption used is that organic matter is approximately 50% (Australian Greenhouse Workbook 1998) and while this may give an error of 14% (based on the actual estimate of 43% C in organic matter), considering many of the other errors in estimation, it may be a reasonable assumption especially when used consistently.

The quantity of carbon dioxide that can be sequestered into biomass is dependent on the type of forest, its rate of growth and the ultimate fate of wood or other harvested biomass. The carbon stored in the ecosystem is found in the vegetation and in the soil. Mature natural forests tend to be in a state of equilibrium with growth being balanced by decay. Such equilibrium may be disturbed, for example by fire, and the loss of carbon dioxide will be followed by a period of growth and net accumulation until a steady state is again reached. Forests which are retained in an essentially undisturbed condition will pass from a period of active growth when carbon dioxide is actively sequestered into biomass, to a period when they mature and where growth and decay are in balance. As trees die and the crowns of large trees become less dense, the amount of dead material on the forest floor can increase substantially and the forest may be a net producer of carbon dioxide. Forests undergoing harvesting will continue to grow and sequester carbon.

The amount of carbon that can be sequestered depends on the rate of growth and the species involved such that the highest rates are in well managed, fast growing species on suitable soils with high water availability. Lower rates of sequestration will occur where rainfall is lower, soils are of lower quality, management is poorer, and inappropriate species are selected.

Information from plantations established in traditional rain-fed areas in the temperate zones provides understanding related to carbon accumulation. Such understanding has relevance to plantations planted in less traditional areas such as those with increasing salinity. Carbon is accumulated in the biomass of the growing trees, both above-ground in stems and branches and below-ground in the roots. The pattern of accumulation typically slows after

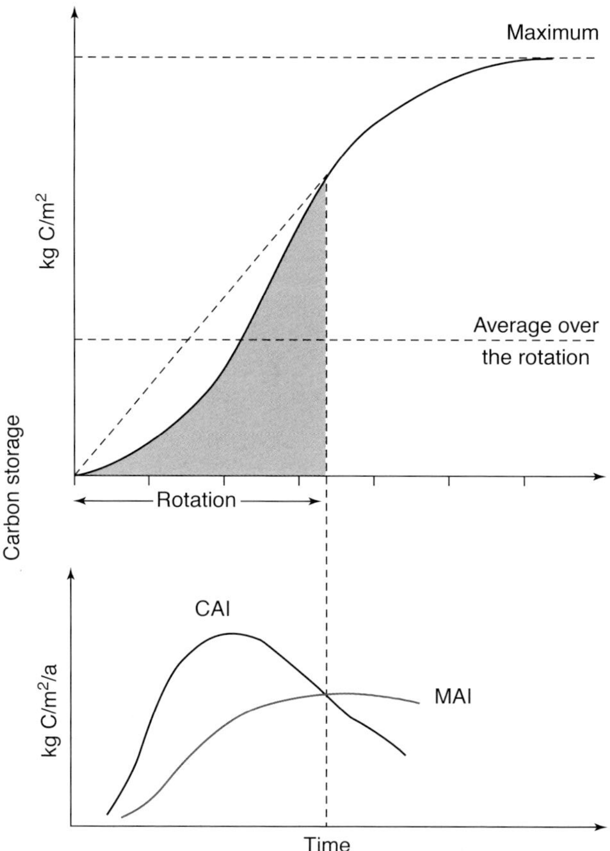

Figure 10.1 The relationship between the point of maximum mean annual increment (MAI) and the amount of carbon stored in a forest (from Cannell and Milne 1995). Maximum MAI occurs when MAI equals the current annual increment (CAI) and is reached while the forest is still accumulating carbon, well before it reaches its maximum carbon content. The amount of carbon stored in a plantation (shaded), when averaged from planting to the point of maximum MAI, is only about one-third of the maximum carbon content that the forest would achieve if it were not harvested.

establishment and this is followed by a period of rapid accumulation after which there is a period of stability. The differences between plantations with different species, different sites and management systems is the length of time where growth is slow, the rate of accumulation in the rapid growth stage, and the level of maximum accumulation. When this is analysed in terms of annual accumulation, there is a peak in productivity (as discussed in Box 4.1). There is a critical relationship between the rotation period to maximum productivity and the actual rotation that is adopted (time to actual harvest). The point of maximum mean annual increment is when the mean annual increment and the current annual increment are equal and is reached when the forest is still accumulating carbon, well before it reaches its maximum carbon content (Figure 10.1). The amount of carbon stored in a plantation, when averaged from time of planting to the point of maximum mean annual

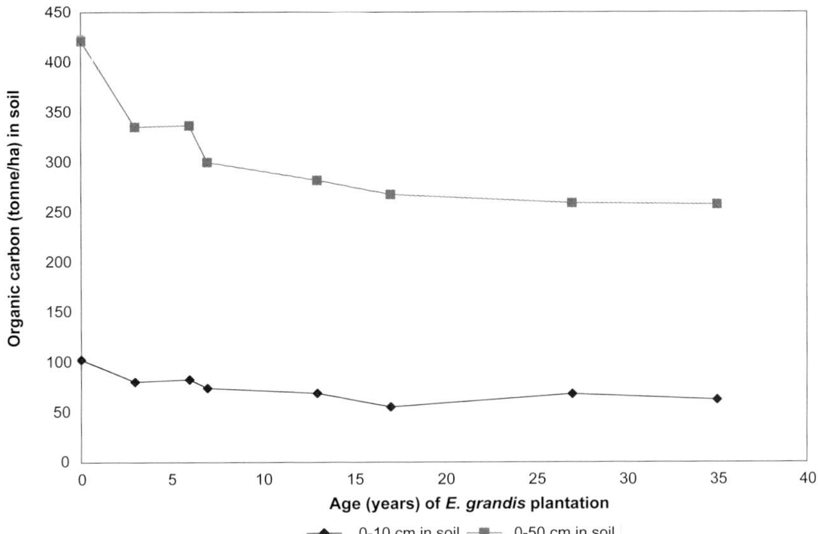

Figure 10.2. Changes in carbon in two soil depths in an *E. grandis* chronosequence in NSW (Turner and Lambert 1999).

increment, is only about one-third of the maximum carbon content that a forest would accumulate if it were not harvested. In management systems designed for carbon accumulation, the objectives are to attempt to maximise carbon accumulation in biomass while minimising carbon losses from the soil.

The result of such an analysis (from Figure 10.1) is that plantations established to accumulate carbon will need to be retained long beyond the period of maximum mean annual increment. The period is probably well beyond the period of optimum return from timber products, that is, the optimum rotation length for carbon accumulation will be different and longer than the optimum rotation length for timber production only. The production graphs have been generalised but the principles for plantations are the same regardless of species or site.

At the time of plantation establishment, the soil is disturbed by ploughing or ripping and there is accelerated organic matter decomposition. The result is a net decline in soil carbon and this will continue until carbon inputs from the developing stand are greater than losses from decomposition. Such a demonstration of soil carbon decline was demonstrated in a *Eucalyptus grandis* plantation (Turner and Lambert 1999). A chronosequence was studied and demonstrated that the surface soil carbon declined in the period immediately after establishment disturbance. After a period of time the quantities were relative stable. However, while there was stability in the surface soil, after a relatively short period of time there were declines in soil carbon at deeper depths and these were maintained for several decades (Figure 10.2). The soils studied were relatively high in carbon and if plantations were established on more degraded sites lower in carbon it would be expected that soil carbon increases would occur at a much more rapid rate.

Within the plantation, a break-even point for carbon occurs where the total carbon accretion in plant biomass equals the total loss from soil organic matter decomposition. Such a break-even point, expressed in years from plantation establishment, will be affected by the depth

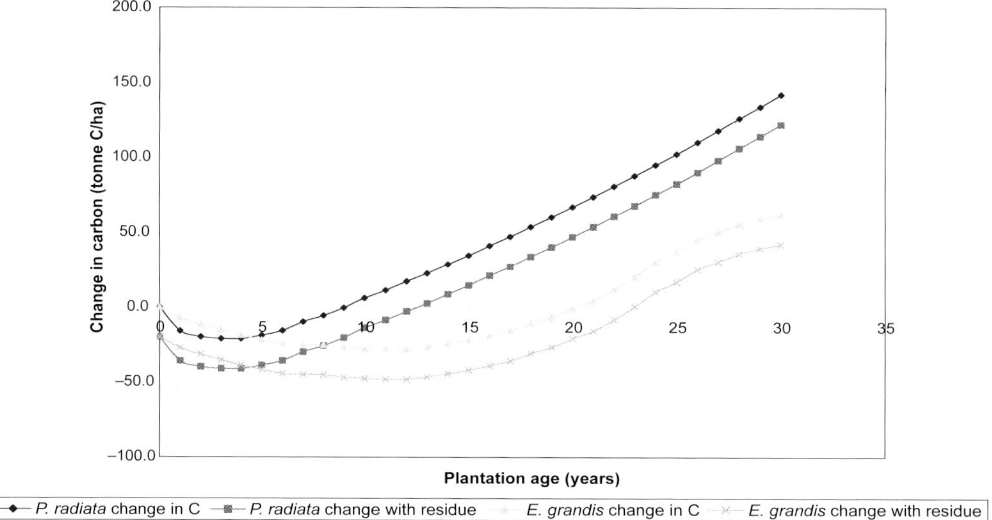

Figure 10.3 Analysis of the break-even point in two plantation chronosequences. The break-even point is where the accumulation in biomass carbon is equivalent to the soil carbon losses. The analyses, one in *E. grandis* and the other in *P. radiata,* also included the effect of loss of carbon from plant residues at the commencement of the plantation (Turner and Lambert 1999).

to which the soil is measured and whether plant residues at time of planting are included in the calculation. Two sites were analysed by the chronosequence method for estimation of break-even point, both on soils relatively high in carbon (Turner and Lambert 1999). One was an *E. grandis* plantation and the other a *P. radiata* plantation. The *P. radiata* plantation took approximately 9 years to break even and 12 years when residues were included, while the *E. grandis* took 20 years and about 22 years when residues were included (Figure 10.3). Where plantations are to be established on saline sites or soils low in carbon and an objective is to increase quantities of carbon stored, it is critical that initial soil disturbance and subsequent silviculture be modified accordingly. However, soils low in carbon at time of planting present a good opportunity for significant carbon accumulation.

The model of carbon accumulation in the total system based on traditional forest areas is not appropriate for areas with increasing salinity and waterlogging. Primarily, they are very low in soil carbon, especially labile carbon, and the effect of site preparation does not necessarily lead to large soil carbon losses. Carbon accumulation in plantations on saline areas will follow patterns similar to more traditional plantations except that the base levels of carbon in the soil may be initially much lower. That is, the tree component of the stand will grow in a similar way but often the levels of carbon in soil at time of establishment will be low to very low, and consequently slow accumulation of carbon in the soil may be a very significant factor. Where silviculture includes rapid growing species, relatively high density of planting and minimised soil disturbance, the result is that establishment of plantations on these areas leads to rapid net accretion of carbon within the system.

Measurement of carbon change in plantations

The objective of a measurement program is to determine the change in carbon storage in the planted area over time. Measurement principles have been described in a number of publications (see Australian Greenhouse Office 1998) and only an outline of requirements is presented here.

The method of estimation of carbon change is to measure the quantity of carbon in the various pools at an initial time and then re-measure at various times in the future, the differences being the carbon accumulation. The initial time is the baseline condition which is critical for estimation of future change. The baseline would preferably be measured prior to any establishment disturbance being undertaken, but when the plantation is already established an alternative is to select adjacent areas which have the same soils and land use prior to plantation establishment. While the location of plots prior to establishment allows for monitoring of change over time, plots in adjacent land use allow an immediate comparison with the plantation. Basic estimations are usually based on the establishment of plots which are then 'scaled up' to the total plantation area.

The elements of a measurement system are:

1 Define and estimate the area of the plantation to be included.

2 Define and estimate areas of individual strata within the total area. The strata will lead to obvious and identifiable variation in carbon accumulation within components including variation in species, age, stand condition, geology, topography and soil type.

3 Locate and establish tree measurement plots within the strata. There are important decisions to be made with regard to plot establishment which include the number to be established, whether they are permanently located to allow re-measurement, and the parameters to be measured in the plot. The number of plots will be determined by the size of the area and the cost of establishment. Suggested approaches have included protocols for one plot every 5 or 10 hectares with a minimum of six plots per strata. The sample plot selection should allow sufficient precision so that samples can be suitably compared with subsequent samplings. The locations of the plots need to be typical of the strata.

4 The plots need to be a suitable size to accommodate changes in stand condition (mortality) such as 30 x 30 metre square plots or circular plots of suitable size. All trees in the plot will be measured for diameter (diameter breast height over bark) and height.

5 If the understorey is significantly developed, it may be measured by destructive sampling on sub-plots, oven dried and weighed.

6 The forest floor to be sampled in sub-plots, oven dried and weighed.

7 Soil can be sampled and analysed as outlined below.

The tree data from the plots can be used for determining biomass of each individual tree, and then by summation of the plot. Individual tree biomass is estimated by two main approaches. In one, a biomass equation, sometimes referred to as an allometric equation, is applied to the tree growth data. Biomass (oven dry kilograms) is the dependent variable in the allometric equation and tree dimensions (diameter and height) are the independent variables. The biomass equations are developed by destructive sampling and estimating

Table 10.2 Examples of published equations for estimating biomass of *E. grandis* based on various independent variables: x = diameter breast height (cm), z = basal area (m^2/ha) at breast height, d = crown depth (m), h = height (m) and y = biomass of component (kg).

Component (y)		Regression parameters			Equation form	R^2	Reference **
		a	*b*	*c*			
Foliage	(kg)	2.9609	0.000898		$y = ax^b$	0.85	1
	(kg)	1.99	1.39		$\log y = a \log x + b$	*	2
Branch	(kg)	3.5072	0.000425		$y = ax^b$	0.87	1
	(kg)	2.59	0.97		$\log y = a \log x + b$	*	2
Bark	(kg)	2.0897	0.0772		$y = ax^b$	0.94	1
	(kg)	3.10	–2.57		$\log y = a \log x + b$	*	2
	(kg/ha)	2.82	0.806	0.619	$\ln y = a + b \ln z + c \ln h$	0.98	3
Sapwood	(kg)	1.9664	0.2838		$y = ax^b$	0.89	1
	(kg)	2.91	–1.81		$\log y = a \log x + b$	*	2
	(kg/ha)	3.59	0.925	0.750	$\ln y = a + b \ln z + c \ln h$	0.99	3
Heartwood	(kg)	3.3626	0.00255		$y = ax^b$	0.94	1
	(kg)	2.75	–1.79		$\log y = a \log x + b$	*	2
Total tree	(kg)	2.3229	0.2214		$y = ax^b$	0.97	1

* Indicates significance at 0.05% level.

** [1] Birk and Turner (1992), [2] Bradstock (1981), [3] Cromer *et al.* (1993)

weight of a selection of trees separate from the measurement plots. Equations may take a variety of forms and include different independent variables. Some published biomass equations for *E. grandis* are given in Table 10.2 primarily showing the differing equation forms. Such equations have been developed for a range of species. Within reasonably acceptable errors, such equations for an individual species covering specified size classes can be applied across a reasonably broad range of stands. That is, once developed, unless errors are apparent, further sampling is not required. When the equations are developed, the biomass of individual components of trees may be assessed, so separate estimates can be made of foliage, branches, bark, wood, total above-ground and roots. Such subdivisions have been made, together with even more refined analyses in some cases, as many of the biomass studies were undertaken for net primary productivity, nutrient cycling or physiological studies. For carbon accretion probably only estimates of the total above-ground and root mass are required.

An alternative method utilises stem volume functions that have been developed for routine production forestry. By applying the diameter and height information to volume functions, the stem volume can be estimated. Two factors are then applied. The first is a density factor which converts the volume to mass and either a generalised factor is used or separate sampling is undertaken of the stem and assessment of wood density. The second factor is to account for the non-stem components (crown, roots) and would have to be obtained from published information or separate sampling as undertaking in the biomass sampling outlined above. The advantage of this approach is that volume estimation is usually undertaken using large numbers of sample trees providing a precise unbiased estimate, whereas the biomass equations are often based on very small numbers of trees over a narrow size class range.

In both cases, the biomass estimate is then converted to carbon content by a conversion factor. A default factor of 0.5 (that is 50% of organic matter is carbon) is often used, but a more precise estimate of 0.43 (43% of organic matter is carbon) is preferable. The estimation of the carbon content in vegetation in a plantation basis is calculated as:

Carbon content of vegetation on a plot basis (tonne C/ha)	=	sum of individual tree above ground biomass *plus* sum of individual tree root biomass *plus* understorey mass *plus* litter mass.
Carbon content of vegetation on a strata basis (tonne C)	=	sum of vegetation carbon content plot basis (in tonnes C/ha) *divided by* number of plots in the strata *multiplied by* area of strata in hectares.
Carbon content of vegetation for the plantation (tonne C)	=	sum of carbon in individual strata.

The errors in the estimates for the plantation and individual strata can be determined from individual plot data using standard statistical methods (Sokal and Rohlf 1981).

The change in carbon over time is estimated by subtracting the carbon estimated at the time of the first measure from the carbon at the second measure, and dividing by the number of years involved. In stands which have not been thinned or harvested, the differences will represent net carbon sequestration. If harvesting has been undertaken to suitably account for change, plot measurements will be needed immediately prior to and after harvesting.

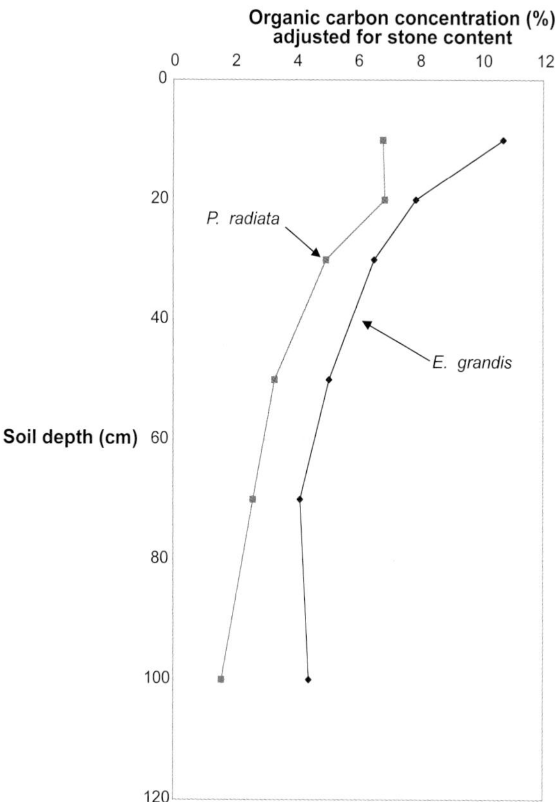

Figure 10.4 Variation in soil carbon with soil depth on two soil types (Turner and Lambert 1999).

Carbon contained within soil can be organic and inorganic and includes finely comminuted carbon, humus, microorganisms, charcoal and some minerals such as carbonates. Mycorhizae and fine roots may be included in the measurement estimation depending on methodology. In many of the traditional forestry areas carbonate is quite rare, but in drier areas where salinity is a problem, bicarbonates and carbonates can be more common and need to be accounted for. Disturbance during establishment of plantations will cause release of soil carbon through the decomposition process and it will be a period of time before the soil carbon returns to its initial level. Soil carbon needs to be measured either prior to establishment to set the baseline or using paired plots to approximate the initial levels.

The estimation of carbon within the soil presents a number of problems mainly due to the high, and not readily predictable, variability of carbon in the soil. This variability is found spatially across the landscape and in soil depth (Figure 10.4). Changes will occur over time and this change will occur at different rates at various soil depths.

Soil sampling may be associated with tree growth plots although soil sampling will usually take considerably more effort. Within a single soil site, the estimation of soil carbon is undertaken for a series of individual depths down to a specified depth. The specified depth has not been universally agreed for carbon calculations but is generally considered as 1 m to 1.5 m.

The individual depths may be specified (for example, 0–50 mm, 50–100 mm, 100–200 mm, 200–500 mm, 500–1000 mm, 1000–1500 mm) or according to horizon. However, sampling by horizon generally requires a much higher level of competence than by specified depths. To calculate the quantity of carbon in a plot to a specified depth:

Soil carbon (tonne C/ha) = soil volume (cc) *multiplied by*
bulk density (g/cc) *multiplied by*
factor for proportion of fine soil (< 2 mm) *multiplied by*
carbon concentration (%).

The type of sampling will be largely determined by the soil type and hence should be planned in conjunction with soil classification and selection of the site for plantation. Usually multiple cores per plot and bulking are undertaken to get a representative sample for chemical analysis. While this will be suitable for many sites, especially for surface soils, it represents a problem for deeper sampling especially where there are significant levels of stone. This may require pit sampling with large block sampling to determine bulk density and rock content.

The estimation of carbon in a sample is usually undertaken on the fine soil component (less than 2 mm) and requires laboratory analysis. In early studies, this type of analysis was undertaken by loss on ignition using a muffle furnace to volatilise organic matter. More recently, the Walkley–Black wet oxidation procedure (Rayment and Higginson 1992) has been undertaken. This method does not directly account for all the carbon in the soil and becomes less accurate with the more recalcitrant carbon found deeper in soil profiles. Experience and development of factors have accounted for this for the more traditional purposes of the analysis in diagnostic studies. Generally, the LECO carbon analyser has been used to estimate total soil carbon. It is critical to understand that different analytical methods will give varying results for a number of reasons, and that for consistency either the same method needs to be maintained or if methods are to be changed during the study period, for example from Walkely–Black to LECO, additional calibration studies need to be undertaken to ensure consistency. The errors introduced may be greater than the actual change in carbon being estimated.

Concluding comments

Carbon accumulation in plantation systems on areas with elevated salinity can be expected to be significant. Such accumulations, especially in the soil, can lead to significant environmental benefits especially related to moisture and nutrient relationships. Generally, periodic assessments will allow the patterns of increase to be ascertained. Carbon potentially may be of value in its own right as part of greenhouse gas reduction strategies, in which case management systems will need to be designed to maximise biomass carbon accumulation and minimise soil carbon losses. In conjunction with this there needs to be an efficient and precise carbon accounting system. Many of the areas with higher salinity suitable for plantation establishment would also be suitable for carbon accumulation because of the low base level of carbon.

Basically, the measurement of carbon is a simple procedure. It is critical to recognise that many traditional forest plantation practices aim to optimise maximum production of specific products and this will probably not be the regime which optimises carbon accumulation. Such a level of compatibility between products requires careful analysis.

REFERENCES

Abrol, I.P., Yadav, J.S.P. and Massoud, F.I. (1988). Salt-affected soils and their management. *FAO Soils Bulletin No. 39*. 131 pp.

Adger, W.N. and Brown, K. (1994). 'Land use and the causes of global warming'. John Wiley and Sons, Chichester. 271 pp.

Ahimana, C. and Maghembe, J.A. (1987). Growth and biomass production by young *Eucalyptus tereticornis* under agroforestry at Morogoro, Tanzania. *For. Ecol. Manage.* **22:** 219–28.

Akilan, K., Farrell, R.C.C., Bell, D.T. and Marshall, J.K. (1997a). Responses of clonal river red gum (*Eucalyptus camaldulensis*) to waterlogging by fresh and salt water. *Aust. J. Exp. Agric.* **37:** 243–8.

Akilan, K., Marshall, J.K., Morgan, A.L., Farrell, R.C.C., Bell, D.T. (1997b). Restoration of catchment water balance: Responses of clonal river red gum (*Eucalyptus camaldulensis*) to waterlogging. *Restoration Ecology* **5:** 101–8.

Allen, J.A., Chambers, J.L. and Stine, M. (1994). Prospects for increasing the salt tolerance of forest trees: a review. *Tree Physiology* **14:** 843–53.

Allender, E.B. (1988). Re-use of municipal effluent and industrial wastewater in eucalypt based afforestation in the River Murray Basin. *In* The International Forestry Conference for the Australian Bicentenary 1988. Vol 2. 18 pp.

Allison, G.B. and Hughes, M.W. (1983). The use of natural tracers as indicators of soil-water movement in a temperate semi-arid region. *J. Hydrol.* **60:** 157-173.

Allison, G.B., Cook, P.G., Barnett, S.R., Walker G.R., Jolly, I.D. and Hughes, M.W. (1990). Land clearance and river salinisation in the western Murray basin. *Aust. J. Hydrology* **119:** 1–20.

Anderson, D.J., Myerscough, P.J. and Pitman, M.G. (1981). Die-back of vegetation in the Central Coast Region. Public Works Department, N.S.W. and Wyong Shire Council, August, Sydney University.

Anon (1980). Quality aspects of farm water supplies. Report of Victorian Irrigation Research and Advisory Services Committee. Published by Victorian Soil Conservation Authority, Victorian Government Printer, Melbourne. 52 pp.

Anon. (1988). Stream salinity issues in Western Australia and approaches to their management. Water Authority of Western Australia. Report No. WS14.

Anon. (1989). Australian Bureau of Statistics, 1989 Agricultural Census (Unpublished results). Australian Government Printing Service, Western Australia.

ANZECC (Australian and New Zealand Environment and Conservation Council) (1992). Australian water quality guidelines for fresh and marine waters. ANZECC, Canberra. 202 pp.

Arunin, S. (1984). Characteristics and management of salt-affected soils in the northeast of Thailand. *In* Ecology and Management of Problem Soils in Asia. Taiwan: Food and Fertilizer Technology Center for the Asian and Pacific Region, Taipai, Taiwan, Republic of China. (FFTC Book Series No. 27). 336–51.

Arunin, S. (1987). Management of saline and alkaline soils in Thailand. Paper presented at the Regional Expert Consultation on the Management of Saline/Alkaline Soils. Bangkok: FAO Regional Office for Asia and the Pacific, 25–29 August 1987. 15 pp.

Astorga, T., Lopez, D., Carazo, N. and Save, R. (1993). Effecto del viento marino en la vegetacion urbana del nuevo litoral Barcelones. *Actas del II Congreso Iberico SECH, Spain,* 539–45.

Aswathappa, N. and Bachelard, E.P. (1986). Ion regulation in the organs of *Casuarina* species differing in salt tolerance. *Aust. J. Plant Physiol.* **13:** 533–45.

Attiwill, P.M. (1980). Nutrient cycling in a *Eucalyptus obliqua* (L'Herit.) forest. IV. Nutrient uptake and nutrient return. *Aust. J. Bot.* **28:** 199–222.

Australian Greenhouse Office (1998). Greenhouse Challenge Vegetation Sinks Workbook. Quantifying carbon sequestration in vegetation management projects. A summary. Prepared by Forestry Technical Services, AACM International and Clean Commodities Inc.

Bacon, P.E., Clatworthy, G. and Nicholls, E. (1993). Tolerance of tree species to wet saline conditions. Report to Catchment Protection Committee, State Forests of NSW, Sydney. 4 pp.

Baker P. (1996). Tree plantations irrigated with chemical factory effluent (background to the Dow Chemical Project). *In* (Eds P.J. Polglase and W.M. Tunningley) 'Land application of wastes in Australia and New Zealand: Research and Practice'. CSIRO, Australia 130–36.

Baker, T., Bail, I., Borschmann, R., Hopmans, P., Morris, J., Neumann, F., Stackpole, D. and Farrow, R. (1994). Commercial tree-growing for land and water care. 2: Trial objectives and design for pilot sites. Produced by the Centre for Tree Technology Victoria, for the Trees for Profit Research Centre, University of Melbourne. 33 pp.

Ballard, T.M. and Cole, D.W. (1974). Transport of nutrients to tree root systems. *Can. J. For. Res.* **4:** 563–65.

Bari, M.A. and Schofield, N.J. (1991). Effects of agroforestry-pasture associations on groundwater level and salinity. *Agroforestry Systems* **16:** 13–31.

Bari, M.A. and Schofield, N.J. (1992). Lowering of a shallow, saline water table by extensive eucalypt reforestation. *J. Hydrol.* **133:** 273–91.

Barker, P.J., Gates, G. and Moore, J.C. (1995). Agroforestry studies for groundwater recharge control near Albury. *In* Proc. Murray Darling 1995 Workshop, Record No. 1995/61. 11–13 September 1995, Wagga Wagga 19–24.

Barrett-Lennard, E.G. (1986). Effects of waterlogging on the growth and NaCl uptake by vascular plants under saline conditions. *Reclamation and Revegetation Research* **5:** 245–61.

Barson, M.M. and Gifford, R.M. (1989). Carbon dioxide sinks: the potential role of tree planting in Australia. *In* (Ed. D.J. Swain). 'Greenhouse and Energy', CSIRO, Melbourne 433–43.

Bartle, J.R. (1995). Sustainable land management with agroforestry. Australian Forestry Council Research Working Group 3 Meeting, Bunbury, Western Australia. (Unpublished report). 12 pp.

Beasley, R. (1994). Electromagnetic induction techniques and its application in north west NSW. *In* Proc. Tamworth Workshop Aust. Soc. Soil Sci. 10–11 March. 99–108.

Bell, D.T. and van der Moezel, P.G. (1988). Stress tolerance in Western Australian *Eucalyptus*: responses to salinity and waterlogging. *In* (di Castri, F., Floret, C., Rambal, S. and Roy, J. eds). Time Scales and Waterstress: Proc 5th Internat. Conference on Mediterranean Systems. The International Union of Biological Sciences, Paris, France 583–8.

Bell, D.T., van der Moezel, P.G., Bennett, I.J., McComb, J.A., Wilkins C.F., Marshall, S.C.B. and Morgan, A.L. (1993). Comparisons of growth of *Eucalyptus camaldulensis* from seeds and tissue culture: root, shoot and leaf morphology of 9-month-old plants grown in deep sand and sand over clay. *For. Ecol. Manage.* **57:** 125–39.

Bell, D.T., McComb, J.A., van der Moezel, P.G., Bennett, I.J. and Kabay, E.D. (1994). Comparisons of selected and cloned plantlets against seedlings for rehabilitation of saline and waterlogged discharge zones in Australian agricultural catchments. *Aust. For.* **57:** 69–75.

Bennett, D.L. and George, R.J. (1995). Using the EM38 to measure the effect of soil salinity on *Eucalyptus globulus* in south-western Australia. *Agricultural Water Management* **27:** 69–86.

Bennett, D.L. and George, R.J. (1996). River red gums (*E. camaldulensis*) for farm forestry on saline and waterlogged land in south-western catchments of Western Australia? *In* Proc. 4th National Conference and Workshop on the Productive Use and Rehabilitation of Saline Lands, Albany,

Western Australia. 25–30 March. Promaco Conventions Pty Ltd, Canning Bridge, Western Australia 345–58.

Benyon, R., Hutchinson, D., Stewart, H.T.L. and O'Shaugnessy, P.J. (1991). Establishment of eucalypt plantations irrigated with sewage at Werribee, Victoria. *In* (Ed. P.J. Ryan) Proc. Third Aust Forest Soils and Nutrition Conference, Productivity in Perspective. Melbourne 7–11 October, Forestry Commission of New South Wales, Sydney 107–8.

Benyon, R.G., Myers, B.J., Theiveyonathon, S., Smith, G.J., Falkiner, R.A., Polglase, P.J. and Bond, W.J. (1996). Effects of salinity on water use, water stress and growth of effluent-irrigated *Eucalyptus grandis*. *In* (Eds P.J. Polglase and W.M. Tunningley) Conference on Land Application of Wastes in Australia and New Zealand: Research and Practice, Canberra 169–74.

Biddiscombe, E.F., Rogers, A.L. and Greenwood, E.A.N. (1981). Establishment and early growth of species in farm plantations near salt seeps. *Aust. J. Ecol.* **6:** 383–9.

Biddiscombe, E.F., Rogers, A.L., Allison, H. and Litchfield, R. (1985a). Response of groundwater levels to rainfall and to leaf growth of farm plantations near salt seeps. *J. Hydrol.* **78:** 19–34.

Biddiscombe, E.F., Rogers, A.L., Greenwood, E.A.N. and de Boer, E.S. (1985b). Growth of tree species near salt seeps, as estimated by leaf area, crown volume and height. *Aust. For. Res.* **15:** 141–54.

Bidner-BarHava, N. and Ramati, B. (1967). The tolerance of some species of *Eucalyptus, Pinus* and other forest trees to soil salinity and low soil moisture in the Negev. *Israel J. Agric. Res.* **17:** 65–76.

Bird, R., Kearney, G.A., Cumming, K.N. and Jowett, D.W. (1993). Effect of tree species and spacing on soil water storage at Hamilton. Proc. Land Management for Dryland Salinity Control, La Trobe University, Bendigo. 28 Sept–1 Oct. Sponsored by the Land and Water Resources Research and Development Corporation.

Birk, E.M. and Turner, J. (1992). Response of flooded gum (*E. grandis*) to intensive cultural treatments: biomass and nutrient content of eucalypt plantations and native forests. *For. Ecol. Manage.* **47:** 1–28.

Blackburn, G. and McLeod, S. (1983). Salinity of atmospheric precipitation in the Murray-Darling Drainage Division, Australia. *Aust. J. Soil Res.* **21:** 411–34.

Blake, T.J. (1981). Salt tolerance of Eucalypt species grown in saline solution culture. *Aust. For. Res.* **11:** 179–83.

Boardman, R. (1991). Impacts on planted trees and soils of re-using effluents and waste waters in Australian environments–a review. *In* (Ed. P.J. Ryan) Proc. Third Aust Forest Soils and Nutrition Conference, Productivity in Perspective. Melbourne 7–11 October, Forestry Commission of NSW, Sydney 87–106.

Boardman, R., Shaw, S., McGuire, D.O. and Ferguson, T. (1996a). Comparison of crop foliage biomass, nutrient contents and nutrient-use efficiency of crop biomass for four species of *Eucalyptus* and *Casuarina glauca* irrigated with secondary sewage effluent from Bolivar, South Australia. *In* (Eds P.J. Polglase and W.M. Tunningley) Conference on Land Application of Wastes in Australia and New Zealand: Research and Practice, Canberra 147–58.

Boardman, R., Shaw, S. and Schrale, G. (1996b). *In* Opportunities and constraints for plantations irrigated with effluent from Adelaide. Proc. Aust. Water & Wastewater Assoc. Conf., Sydney, May, 287–94.

Boden, D.I. (1990). The relationship between soil water status, rainfall and the growth of *Eucalyptus grandis*. Paper presented at 19th IUFRO World Congress, August, Montreal, Canada. International Union of Forest Research Organisations, Vienna, Austria.

Boland, D.J., Brophy, J.J. and House, A.P.N. (1991). *Eucalyptus* Leaf Oils: use, chemistry, distillation and marketing. Inkata Press, Melbourne, Australia. 252 pp.

Boland, D.J., Brooker, M.I.H., Chippendale, G.M., Hall, N., Hyland, B.P.M., Johnston, D.A., Kleinig, D.A. and Turner, J.D. (1992). 'Forest Trees of Australia'. CSIRO Publications, Melbourne, Australia. 687 pp.

Bonny, L. (1991). Growth of a *Eucalyptus grandis* plantation following intensive silvicultural treatments applied in the first six years. Forestry Commission of New South Wales Research Paper No. 12. 19 pp.

Borg, H., Stoneman, G.L. and Ward, C.G. (1987). Steam and ground water response to logging and subsequent regeneration in the southern forest of Western Australia. Results from four catchments. Department of Conservation and Land Management, Manjimup, Western Australia, Technical Report No. 16. 103 pp.

Borg, H., Bell, R.W. and Loh, I.C. (1988). Streamflow and stream salinity in a small water supply catchment in southeast Western Australia after reforestation. *J. Hydrol.* **103:** 323–33.

Boyce, S.G. (1954). The salt spray community. *Ecological Monograph* **24:** 29–67.

Bradstock, R. (1981). Biomass in an age series of *Eucalyptus grandis* plantations. *Aust. For. Res.* **11:** 111–27.

Bren, L., Hopmans, P., Gill, B., Baker, T. and Stackpole, D. (1993). Commercial tree-growing for land and water care. 1: Soil and groundwater characteristics of the pilot sites. A Report to the Trees for Profit Research Board. Produced by the Centre for Forest Tree Technology for the Trees for Profit Research Centre, University of Melbourne. 46 pp.

Buchanan, A.H. and Honey, B.G. (1994). Energy and carbon dioxide implications of building construction. *Energy and building* **20:** 205–217.

Buckley, G.P. (1988). Soil factors influencing yields of *Eucalyptus camaldulensis* on former tin-mining land in the Jos Plateau region, Nigeria. *For. Ecol. Manage.* **23;** 1–17.

Burch, G., Graetz, D. and Noble, I. (1987). Biological and physical phenomena in land degradation. *In* (Eds A. Chisholm and R. Dumsday) 'Land Degradation: Problems and Policies' Cambridge University Press, U.K 27–48.

Bussotti, F., Grossoni, P. and Pantani, F. (1995). The role of marine salt and surfactants in the decline of Tyrrhenian coastal vegetation in Italy. *Ann. Sci. For.* **52:** 251–61.

Butcher, T.B. (1983). Variation in salt tolerance of Wandoo. *In* (Ed. P. Cotterill) Proceedings of the 8th Meeting of Research Working Group No. 1, Forest Genetics, 14–19 November, Perth, Western Australia. Standing Committee of Australian Forestry Council, Canberra. 114 pp.

Cameron, D.R., de Jong, E., Read, D.W.L. and Oosterveld, M.I. (1981). Mapping salinity using resistivity and electromagnetic induction techniques. *Can. J. Soil Sci.* **61:** 67–78.

Campinhos, E. (1980). More wood of better quality through intensive silviculture with rapid-growth improved Brazilian Eucalyptus. *Tappi* **63:** 145–47.

Cannell, M.G.R. and Milne, R. (1995). Carbon pools and sequestration in forest ecosystems in Britain. *Forestry* **68:** 361–78.

Carbon, B.A., Bartle, G.A. and Murray, A.M. (1981). Water stress, transpiration and leaf area index in eucalypt plantations in a bauxite mining area in south-west Australia. *Aust. J. Ecol.* **6**: 459–66.

Carron, L. T. (1990). A history of plantation policy in Australia. *In* (Eds. J. Dargavel and N. Semple) Prospects for Australian Forest Plantations. Centre for Resource and Environmental Studies, Australian National University, Canberra, Australia. 11–23.

Cassidy, N.G. (1968). The effect of cyclic salt in a maritime environment. II. The absorption by plants of colloidal atmospheric salt. *Plant Soil* **28:** 390–406.

Charman, P.E.V. and Murphy, (1991). Soils, their properties and management. 'A Soil Conservation Handbook for NSW'. Sydney University Press. 363 pp.

Charman, P.E.V. and Junor, R.S. (1989). Saline seepage and land degradation–a New South Wales perspective. BMR *Journal of Australian Geology & Geophysics* **11(2/3):** 195–203.

Chaturvedi, A.N., Sharma, S.C. and Srivastava, R. (1984). Water consumption and biomass production of some forest areas. *Common. For. Rev.* **63:** 217–23.

Chaudary, M.B., Mian, M.A. and Rafiq, M. (1978). Nature and magnitude of salinity and drainage problems in relation to agricultural development in Pakistan. *Pakistan Soils Bulletin No. 8.*

Clemens, J. and Pearson, C.J. (1977). The effect of waterlogging on the growth and ethylene content of *Eucalyptus robusta* Sm. (Swamp Mahogany). *Oecologia* **29:** 249–55.

Clemens, J., Kirk, A.M. and Mills, P.D. (1978). The resistance to waterlogging of three *Eucalyptus* species: effect of waterlogging and an ethylene-releasing growth substance on *E. robusta, E. grandis* and *E. saligna*. *Oecologia* 34: 125-131.

Clemens, J., Campbell, L.C. and Nurisjah, S. (1983). Germination, growth and mineral ion concentrations of *Casuarina* species under saline conditions. *Aust. J. Bot.* **31:** 1–9.

Clifton, C.A. (1992). Tree growth and crown development in plantations established to control groundwater recharge. Department of Conservation and Environment, Centre for Land Protection Research, Technical Paper No. 4. 20 pp.

Conacher, A.J. (1975). Throughflow as a mechanism responsible for excessive soil salinisation in non-irrigated, previously arable lands in Western Australian wheatbelt: A field study. *Catena* **2:** 31–67.

Connell, L., Morris, J., Heuperman, A., Vertessey, R., Feckema, P., Silberstein, R., Mann, L., Collopy, J. and Komarzynski, M. (1997). The impact of dryland agriculture on land and river salinization in south-western New South Wales. *Aust. J. Soil Water Conservation* **10:** 29–36.

Cook, P.G. and Walker, G.R. (1989). Groundwater recharge in south-western New South Wales. Centre for Research in Groundwater Processes. Report No. 9. 31 pp.

Cook, P.G., Kennett-Smith, A., Walker, G.R., Budd, G., Williams, M. and Anderson, R. (1997). The impact of dryland agriculture on land and river salinisation in south-western New South Wales. *Aust. J. Soil Water Conservation* **10:** 29–36.

Corwin, D.L. and Rhoades, J.D. (1982). An improved technique for determining soil electrical conductivity–depth relations from aboveground measurement. *Soil Sci. Soc. Am. J.* **46:** 517–20.

Craig, G.F., Bell, D.T. and Atkins, C.A. (1991). Response to salt and waterlogging stress of ten taxa of *Acacia* selected from naturally saline areas of Western Australia. *Aust. J. Bot.* **38:** 619–30.

Cromer, R.N., Cameron, D.M., Rance, S.J., Ryan, P.A. and Brown, M. (1993). Response to nutrients in *Eucalyptus grandis*. 1. Biomass accumulation. *For. Ecol. Manage.* **62:** 211–30.

Cromer, R.N., Cameron, D.M., Lennon, S., Rance, S.J., Ryan, P.A. and West, P.W. (1995). Response of *Eucalyptus grandis* to fertilizer and irrigation, to five years of age. Contribution to Australian Forestry Council, Research Working Group 3 (Soils and Nutrition) Meeting at Busselton, Western Australia, 25–9 September. 5 pp.

Czarnowski, M.S., Humphreys, F.R. and Gentle, S.W. (1971). Quantitative expression of site index in terms of certain soil and climate characteristics of *Pinus radiata* (D. Don) plantations in Australia and New Zealand. *Ekologia Polska* **19:** 295–309.

Dale, M. (1992). Preliminary results on irrigation drainage water re-use on different tree species. *In* Catchments of Green, Adelaide, South Australia. Proc. Murray–Darling Basin Division National Workshop. Vol. B p. 177–82.

Davison, A.W. (1971). The effects of deicing salts on roadside verges. I. Soil and plant analyses. *J. Appl. Ecol.* **8:** 555–561.

Davison, E.M. and Tay, F.C.S. (1985). The effect of waterlogging on seedlings of *Eucalyptus marginata*. *New Phytologist* **101:** 743–53.

Dell, B., Malajczuk, N. and Grove, T.S. (1995). Nutrient disorders in plantation eucalypts. ACIAR Monograph 31. 110 pp.

Dirr, M.A. (1974). Tolerance of honeylocust seedlings to soil applied salt. *Hort Science* **9:** 53–4.

Dirr, M.A. (1976). Selection of trees for tolerance to salt injury. *J. Arbor.* **2:** 209–16.

Dong, A., Tanji, K., Grattan, S., Kavanagh, F. and Parlange, M. (1992). Water quality effects on *Eucalyptus* E.T. Irrigation and Drainage Proceedings, Water Forum 92. ASCE Baltimore, MD. August 1992.

Doran, J.C. and Brophy, J.J. (1990). Tropical red gums–a source of 1,8-cineole-rich *Eucalyptus* oil. *New Forests* **4:** 157–78.

Doran, J.C. and Matheson, A.C. (1994). Genetic parameters and expected gains from selection for monoterpene yields in Petford *Eucalyptus camaldulensis*. *New Forests* **8:** 155–67.

Dowden, H.G.M., Lambert, M.J. and Truman, R.A. (1978). Salinity damage to Norfolk Island pines caused by surfactants. II. Effects of sea-water and surfactant mixtures on the health of whole plants. *Aust. J. Plant Physiol.* **5:** 387–95.

Dowden, H.G.M. and Lambert, M.J. (1979). Environmental factors associated with a disorder affecting tree species on the coast of New South Wales with particular reference to Norfolk Island pines *Araucaria heterophylla*. *Environmental Pollution* **19:** 71–84.

Dudal, R. and Purnell, M.F. (1986). Land resources: salt affected soils. *Reclamation Revegetation Research* **5:** 1010.

Dumsday, R.G., Pegler, R. and Oram, D.A. (1989). Is broadscale revegetation economic and practical as a groundwater and salinity management tool in the Murray–Darling Basin? *BMR Journal of Australian Geology & Geophysics* **11:** 209–18.

Dumsday, R.G. and Oram, D.A.. (1990). Economics of dryland salinity control in the Murray River Basin, Northern Victoria (Australia). *In* (Eds J.A. Dixon, D.E. James and P.B. Sherman) 'Dryland Management: Economic Case Studies'. Earthscan Publications Ltd, London, U.K 215–240.

Duncan, M.J., Baker, T.G., Stackpole, D.J. and Hamlet, A.G. (1997). Monitoring of irrigation water quality, chemical properties of surface soil and tree nutrition, at the Trees for Profit Pilot Sites 1992–96. Centre for Forest Tree Technology, Victoria. 15 pp.

Dye, P.J. (1979). Feasibility of reclaiming and improving sodic soils. *Rhodesia agric. J.* **76:** 65–8.

Dye P.J. and Poulter, A.G. (1995). A field demonstration of the affect on streamflow of clearing invasive pine and wattle trees from a riparian zone. *South African Forestry Journal* **173:** 27–36.

Dyson, P.R. (1983). Dryland salting and groundwater discharge in the Victorian uplands. *Proc. Royal Soc. Victoria* **95:** 113–6.

Eastham, J., Scott, P.R., Steckis, R.A., Barton, A.F.M., Hunter R.J. and Sudmeyer, R.J. (1993). Survival, growth and productivity of tree species under agroforestry to control salinity in the Western Australian wheatbelt. *Agroforestry Systems* **21:** 223–37.

Edgar, J.G and Stewart, H.T.L. (1979). Wastewater disposal and reclamation using *Eucalyptus* and other trees. *Prog. Wat. Tech.* **11:** 163–73.

Edlin, H.L. (1957). Saltburn following a summer gale in south-east England. *Quarterly J. Forestry* **51:** 46–50.

Edwards, G.A., Fish, N.W., Fuell, K.J., Keil, M., Purse, J.G. and Wignall, T.A. (1995). Genetic modifications of eucalypts – objectives, strategies and progress. *In* Theme 6. Emerging technologies. 'Eucalypt Plantations: Improving Fibre Yield and Quality'. Cooperative Research Centre on Temperate Hardwood Plantations, Hobart, Tasmania, Australia 389 -91.

Edwards, J. and Harper, R. (1996). Site evaluation for *Eucalyptus globulus* in south-western Australia for improved productivity and water uptake. *In* Proc. 4th National Conference and Workshop on the Productive Use and Rehabilitation of Saline Lands, Albany, Western Australia. 25–30 March. Promaco Conventions Pty Ltd, Canning Bridge, Western Australia 197–203.

Eldridge, K. Davidson, J., Harwood, C. and van Wyk, G. (1993). 'Eucalypt domestication and breeding'. Clarendon Press, Oxford. 288 pp.

El-lakany, M.H. and Luard, E.J. (1982). Comparative salt tolerance of selected *Casuarina* species. *Aust. For. Res.* **13:** 11–20.

Engel R. (1987). Agroforestry management of saltland using salt tolerant eucalypts. *Trees and Natural Resources* **28:** 12–3.

Engel, R. and Negus, T. (1988). Controlling saltland with trees. Western Australian Department of Agriculture, Farmnote 46/88, Agdex 331/570.

Epstein, E. (1972). 'Mineral nutrition of plants: Principles and perspectives'. J. Wiley and Sons, Inc., New York. 412 pp.

Falkiner, R.A. and Smith, C.J. (1997). Changes in soil chemistry in effluent-irrigated *Pinus radiata* and *Eucalyptus grandis* plantations. *Aust. J. Soil Res.* **35:** 131–47.

Farrington, P. and Salama, R.B. (1996). Controlling dryland salinity by planting trees in the best hydrogeological setting. *Land Degradation and Development* **7:** 183–204.

Farrell, R.C.C., Bell, D.T., Akilan, K. and Marshall, J.K. (1996a). Morphological and physiological comparisons of clonal lines of *Eucalyptus camaldulensis*. 1. Responses to drought and waterlogging. *Aust. J. Plant Physiol.* **23:** 497–507.

Farrell, R.C.C., Bell, D.T., Akilan, K. and Marshall, J.K. (1996b). Morphological and physiological comparisons of clonal lines of *Eucalyptus camaldulensis*. 2. Responses to waterlogging salinity and alkalinity. *Aust. J. Plant Physiol.* **23:** 509–18.

Feikema, P.M. and Connell, L.D. (1995). Shallow groundwater movement in response to eucalypt plantation water use. p. 285. *In* Murray Darling Basin 1995 Workshop, Wagga Wagga, 11–13 September, Research Paper 1995/61, Murray Darling Basin Commission.

Felker, P., Clark, P.R., Laag, A.E. and Pratt, P.F. (1981). Salinity tolerance of the tree legumes: mesquite (*Prosopis glandulosa* var. *torreyana*, *P. velutina* and *P. articulata*), and algarrobo (*P. chilensis*), kiawe (*P. pallida*) and tamarugo (*P. tamatugo*) grown in sand culture on nitrogen-free media. *Plant and Soil* **61:** 311–7.

Firmin, R. (1968). Forestry trials with highly saline or sea water in Kuwait. *In* (Ed. H. Boyles) Saline irrigation for Agriculture and Forestry. Junk, The Hague. 325 pp.

Fitzpatrick, D. (1994). 'Money Trees on Your Property'. Inkata Press. p. 20–8.

Flinn, D.W., Bren L.J. and Hopmans, P. (1979). Soluble nutrient inputs from rain and outputs in stream water from small forested catchments. *Aust. For.* **42:** 39–49.

Flowers, T.J., Troke, P.F. and Yeo, A.R. (1977). The mechanism of salt tolerance in halophytes. *Ann. Rev. Plant Physiol.* **28:** 89–121.

Foster, R.C. and Sands, R. (1977). Response of radiata pine to salt stress. II. Localisation of chloride. *Aust. J. Plant Physiol.* **4:** 863–75.

Fox, J.E.D., Neilsen, J.R. and Osborne, J.M. (1990). *Eucalyptus* seedling growth and salt tolerance from the north-eastern goldfields of Western Australia. *J. Arid Environ.* 19: 45-53.

Fraser, G.W., Thorburn, P.J. and Taylor, D.W. (1996). Growth and water use of trees in saline land in south east Queensland. *In* Proc. 4th National Conference and Workshop on the Productive Use and Rehabilitation of Saline Lands, Albany, Western Australia. 25–30 March. Promaco Conventions Pty Ltd, Canning Bridge, Western Australia 219–24.

Gafni, A. (1997). Irrigation and the Environment: An Israeli perspective. Paper presented at ANCID Conference 'Striking the Balanc', Deniliquin, NSW.

Gasana, J.K. and Loewenstein, H. (1984). Site classification for Maiden's gum (*Eucalyptus globulus* subsp. *maidenii*) in Rwanda. *For. Ecol. Manage.* **8:** 107–16.

Gellini, F., Pantani, F., Grossoni, P., Bussotti, F., Barbolani, E. and Rinallo, C. (1985). Further investigation on the causes of disorder of the coastal vegetation in the park of San Rossore (central Italy). *European Journal of Forest Pathology* **15:** 145–57.

George, R.J. (1990). Reclaiming sandplain seeps by intercepting perched groundwater with eucalypts. *Land Degradation and Rehabilitation* **2:** 13–25.

Ghassemi, F., Jakeman, A.J. and Nix, H.A. (1995). Salinisation of Land and Water Resources: Human causes, extent, management and case studies. University of New South Wales Press Ltd, Sydney, Australia. 526 pp.

Gill, H.S. and Abrol, I.P. (1991). Salt affected soils, their afforestation and its ameliorating influence. *The International Tree Crops Journal* **6:** 239–60.

Goldsmith, E. and Hildyard, N. (1988). Battle for the earth: today's key environmental issues. Child and Associates, Brookvale, Australia. 240 pp.

Grattan, S.R., Shannon, M.C., Grieve, C.M., Poss, J.A., Suarez, D. and Leland, F. (1997). Interactive effects of salinity and boron on the performance and water use of *Eucalyptus. In* (Ed. K.S. Chartzoulakis) Proc. 2nd Int. Symp. on Irrigation of Horticultural Crops. *Acta Hort.* 449 Vol **2** 607–13.

Greenway, H. (1965). Plant response to saline substrates. VII. Growth and ion uptake throughout plant development in two varieties of *Hordeum vulgare. Aust. J. Biol. Sci.* **18:** 763–79.

Greenway, H. and Munns, R. (1980). Mechanisms of salt tolerance in nonhalophytes. *Ann. Rev. Plant Physiology* **31:** 149–90.

Greenwood, E.A.N. (1986). Water use by trees and shrubs for lowering saline groundwater. *Reclamation and Revegetation Research* **5:** 423–34.

Greenwood, E.A.N. (1992). Deforestation, revegetation, water balance and climate: an optimistic path through the plausible, impracticable and the controversial. *Advances in Bioclimatology* **11:** 89–154.

Greenwood, E.A.N., Klein, L., Beresford, J.D. and Watson, G.D. (1985). Differences in annual evaporation between grazed pasture and *Eucalyptus* species in plantations on a saline farm catchment. *J. Hydrol.* **78:** 261–78.

Greenwood, E.A.N., Milligan, A., Biddiscombe, E.F., Rogers, A.L., Beresford, J.D., Watson, G.D. and Wright, K.D. (1992). Hydrologic and salinity changes associated with tree plantations in a saline agricultural catchment in south-western Australia. *Agric. Water. Manage.* **22:** 307–23.

Greenwood, E.A.N., Biddiscombe, E.F., Rogers, A.L., Beresford J.D. and Watson, G.D. (1994). The influence of groundwater levels and salinity of a multi-species tree plantation in the 500 mm rainfall region of south-eastern Australia. *Agric. Water. Manage.* **25:** 179–84.

Greenwood, E.A.N., Biddiscombe, E.F., Rogers, A.L., Beresford J.D. and Watson, G.D. (1995). Growth of species in a tree plantation and its influence on salinity and groundwater in the 400 mm rainfall region of south-western Australia. *Agric. Water. Manage.* **28:** 231–43.

Grewal, S.S. and Abrol, I.P. (1986). Agroforestry on alkali soils: effects of some management practices on initial growth, biomass accumulation and chemical composition of selected species. *Agroforestry Systems* **4:** 221–32.

Grieve, A.M. and Pitman, M.G. (1978). Salinity damage to Norfolk Island pines caused by surfactants. III. Evidence for stomatal penetration as the pathway of salt entry to leaves. *Aust. J. Plant Physiol.* **5:** 397–413.

Gutteridge, Haskins and Davey (1970). Murray Valley Salinity Investigation. A report to the River Murray Commission, Canberra, Australia. 67 pp.

Hafeez, S. M. (1993). Identification of fast growing salt tolerant tree species. *Pakistan J. Forestry* **43:** 216–20.

Hamlet, A. and Morris, J. (1996). Effects of salinity on the growth of four *Eucalyptus* species on irrigated sites in Northern Victoria. A Report to the Trees for Profit Research Board. Produced by the Centre for Forest Tree Technology for the Trees for Profit Research Centre, University of Melbourne. 16 pp.

Hanna, D., Shaw, S., Boardman, R. and Schrale, G. (1992). Early performance of effluent irrigated hardwood plantations at Bolivar, South Australia. Volume B. *In* Proc. Conference on Catchments of Green, Greening Australia, Adelaide. 23–26 March, 1992. 207–14.

Hart, B.T. and McKelvie, I.D. (1986). Chemical limnology in Australia. *In* (Eds. P. de Decker and W.D. Williams) Limnology in Australia. CSIRO, Melbourne, Australia. pp. 3–32.

Hasegawa, P.M., Bressan, R.A. and Handa, A.K. (1986). Cellular mechanisms of salinity tolerance. *Hort. Sci.* **21:** 1317–24.

Hatton, T.J., Pierce, L.L. and Walker, J. (1993). Ecohydrological changes in the Murray–Darling Basin. II. Development and tests of a water balance model. *J. Appl. Ecol.* **30:** 274–82.

Heath, J. and Heuperman, A. (1994). Serial biological concentration of salts; its application to the Shepparton irrigation region. *In* (Eds. M.S. Schulz and G. Petterson). Productive Use of Saline Lands. Saline Irrigation Areas. Proceedings of a Workshop, Echuca, Australia. 15–17 March 1994. 95–103.

Herbert, M.A. and Schonau, A.P.G. (1990). Fertilizing commercial forest species in southern Africa: research progress and problems. Part 2. *S. For. J.* **152:** 34–42.

Hillel, D. (1990). Ecological aspects of land drainage for salinity control in arid and semi-arid regions. *In* Symposium on Land Drainage for Salinity Control in Arid and Semi-arid Regions, Cairo, Egypt. *Drainage Research Institute* **1:** 125–35.

Hingston, F.J. and Gailitis, V. (1976). The geographic variation of salt precipitated over Western Australia. *Aust. J. Soil Res.* **14:** 319–35.

Hingston, F.J. and Galbraith, J.H. (1998). Application of the process-based model BIOMASS to *Eucalyptus globulus* subsp. *globulus* plantations on ex-farmland in south-western Australia. 2. Stemwood production and seasonal growth. *For. Ecol. Manage.* **106:** 157–68.

Hingston, F.J., Galbraith, J.H. and Dimmock, G.M. (1998). Applications of the process-based model BIOMASS to *Eucalyptus globulus* subsp. *globulus* plantations on ex-farmland in south-western Australia. 1. Water use by trees and assessing risk of losses due to drought. *For. Ecol. Manage.* **106:** 141–56.

Holmes, F.W. (1961). Salt injury to trees. *Phytopathology* **5:** 712–8.

Holmes, F.W. (1966). Salt injury to trees. II. Sodium and chloride in roadside sugar maples in Massachusetts. *Phytopathology* **56:** 533–6.

Hookey, G.R. (1987). Prediction of delays in groundwater response to catchment clearing. *J. Hydrol.* **94:** 181–98.

Hopmans, P. Stewart, H.T.L., Flinn, D.W. and Hillman, T.J. (1990). Growth, biomass production and nutrient uptake by seven tree species irrigated with municipal effluent at Wodonga, Australia. *For. Ecol. Manage.* **30:** 203–11.

House of Representatives, Standing Committee on Environment, Recreation and the Arts. (1989).

Hoy, N.T., Gale, M.J. and Walsh, K.B. (1994). Revegetation of a scalded discharge zone in central Queensland. I. Selection of tree species and evaluation of an establishment technique. *Aust. J. Experimental Agric.* **34:** 765–76.

Hussain, A. and Gul, P. (1992). Selection of tree species for saline and waterlogged areas in Pakistan. *In* (Eds. N. Davidson and R. Galloway). Productive use of salt-affected land. ACIAR Proceedings No. 42, Canberra, Australia. 53-55.

Hussain, A.M.M. and Theagarajan, K.S. (1966). Preliminary studies in the mineral nutrition of *Eucalyptus* 'hybrid' seedlings. *Indian Forester* **92:** 285–92.

Inions, G. (1991). Relationships between environmental attributes and the productivity of *Eucalyptus globulus* in south-west Western Australia. *In* (Ed. P.J. Ryan) Proc. Third Aust Forest Soils and Nutrition Conference, Productivity in Perspective. Melbourne 7–11 October, Forestry Commission of NSW, Sydney. 116–32.

IPCC (Inter Governmental Panel on Climate Change) (1996). Gas Inventories: Revised Reference Manual. Chapter 5: Land-Use Change and Forestry. Cambridge University Press, Cambridge.

Jacobsen, T. and Adams, R.M. (1958). Salt and silt in ancient Mesopotamian agriculture. *Science* **128:** 1251–8.

Jain, B.L., Muthana, K.D. and Goyal, R.S. (1985). Performance of tree species in salt-affected soils in arid regions. *J. Indian Soc. Soil Sci.* **33:** 221–4.

Jenkin, J.J. (1981). Terrain, groundwater and secondary salinity in Victoria, Australia. *Agric. Water Management* **4:** 143–72.

Jenny, H. (1980). 'The soil resource, origin and behaviour'. Springer-Verlag, New York. 377 pp.

Johnson, I.G. and Stanton, R.R. (1993). Thirty years of eucalypt species and provenance trials in New South Wales–survival and growth trials established from 1961 to 1990. *Forestry Commission of New South Wales Research Paper No. 20*. 92 pp.

Johnstone, P.C. (1984). Chemosystematics and morphometric relationships in populations of some peppermint eucalypts. PhD Thesis, Monash University, Victoria, Australia.

Kabay, E.D., Thomson, L.A.J., Doran, J.C., van der Moezel, P.J., Bell, D.T., McComb, J.A., Bennett, I.J., Hartney, V.J. and Malajczuk, N. (1986). Micropropogation of forest trees selected for salt tolerance. *Aust. Salinity Newsletter* **14:** 60–2.

Kaddah, M.T. and Rhoades, J.D. (1976). Salt and water balance in Imperial Valley, California. *Soil Sci. Soc. Am. J.* **40:** 93–100.

Kafkafi, U. (1984). Plant nutrition under saline conditions. *In* (Eds. I. Shainberg and J. Shalhevet) 'Soil salinity under irrigation'. Springer-Verlag, Berlin. 319–38.

Karajeh, F.F. and Tanji, K.K. (1994a). Agroforestry drainage management model. II. Field water flow. *J. Irrig. Drainage Engineering ASCE*, **120:** 382–96.

Karajeh, F.F. and Tanji, K.K. (1994b). Agroforestry drainage management model. III. Field salt flow. *J. Irrig. Drainage Engineering ASCE*, **120:** 397–413.

Karschon, R. (1958). Leaf absorption of wind-borne salt and leaf scorch in *Eucalyptus camaldulensis* Dehn. *Ilanoth* **4:** 5–25.

Karschon, R. (1963a). A probable case of magnesium deficiency in *Eucalyptus gomphocephala* A.Dc. *Israel J. Agric. Res.* **13:** 173–5.

Karschon, R. (1963b). Key to leaf symptoms of mineral deficiencies in *Eucalyptus gomphocephala* A.Dc. *Israel J. Agric. Res.* **13:** 177–81.

Karschon, R. (1966). Environment and tree growth in the Rift Valley Oasis of Ein Hatseva. *La-Yaaran Supplement* 2.

Karschon, R. and Heth, D. (1967). The water balance of a plantation of *Eucalyptus camaldulensis* Dehn. *In* 'Contribution on Eucalypts in Israel'. III, 'Ahva' Coop. Press, Jerusalem. 7–34.

Karschon, R. and Zohar, Y. (1975). Effects of flooding and of irrigation water salinity on *Eucalyptus camaldulensis* Dehn. from three seed sources. Leaflet 54. Agricultural Research Organisation, Ilanot, Israel. 7 pp.

Kennedy, A. (1992). Irrigated winery woodlot case study. *In* Catchments of Green, Adelaide, South Australia. Proc. Murray–Darling Basin Division National Workshop. Vol. B. 197–206.

Kennett-Smith, A.K., Budd, G.R., Cook, P.G. and Walker, G.R. (1990). The effect of lucerne on the recharge to cleared mallee lands. Centre for Research in Groundwater Processes, Report No. 27. 40 pp.

Kirschbaum, M.U.F., Fischlin, A., Cannell, M.G.R., Cruz, R.V., Galinski W., and Cramer, W.A. (1996). Climate change impacts on forests. *In* (Eds R.T. Watson, M.C. Zinyowera and R.H. Moss). Climate Change 1995: Impacts, adaptations, and mitigation of climate change: Scientific–Technical analyses. Contribution of Working Group II to the second assessment report of the Intergovernmental Panel on Climate change. Cambridge University Press, Cambridge. 131–58.

Knott, J. (1996). Financial analysis of the use of cloned, superior genotypes in eucalypt plantations. State Forests of New South Wales, Internal Report.

Knott, J. and Turner, J. (1990). Fertilizer usage in Forestry Commission of New South Wales exotic conifer plantations, 1955–89. Forestry Commission of New South Wales Technical Paper No. 51. 95 pp.

Kramer, P.J. (1969). 'Plant and soil water relationships: a modern synthesis'. McGraw-Hill Book Co. New York. 482 pp.

Kube, P.D., Sandell, P. and Rose, B. (1987). Growth of *Eucalyptus camaldulensis* irrigated with sewage effluent in central Australia. Conservation Commission of the Northern Territory, Technical Memorandum 87/5. 23 pp.

Kurauchi, I. (1956). Salt spray damage to the coastal forests. *Japanese J. Ecology* **5:** 13–7.

Lacaze, J.F. (1978). Etude de l'adaption ecologique des eucalyptus, etude de provenances d'Eucalyptus camaldulensis project. FAO Project No. 6. *In* FAO Third World Consultation on Forest Tree Breeding. 21–26 March, 1977. Vol 1. CSIRO, Canberra, Australia. 393–409.

Ladiges, P.Y. and Kelso, A. (1977). The comparative effects of waterlogging on two populations of *Eucalyptus viminalis* Labill. and one population of *E. ovata* Labill. *Aust. J. Bot.* **25:** 159–69.

Lambert, M.J. (1981). Inorganic constituents in wood and bark of New South Wales forest tree species. *Forestry Commission of New South Wales Research Note No. 45*. 43 pp.

Larcher, W. (1980). 'Physiological Plant Ecology'. Springer Verlag, Berlin. 303 pp.

Lassak, E.V. (1988). The Australian *Eucalyptus* oil industry, past and present. *Chemistry in Australia* **55:** 396–406.

Lesch, S.M., Rhoades, J.D., Lund, L.J. and Corwin, D.L. (1992). Mapping soil salinity using calibrated electromagnetic measurements. *Soil Sci. Soc. Am. J.* **56:** 540–8.

Letey, J. and Knapp, K.C. (1995). Simulating saline water management strategies with application to arid-region agroforestry. *J. Environ. Qual.* **224:** 934–40.

Loh, I.C. (1985). Stream salinity control by increased transpiration of vegetation. Water Authority of Western Australia. Report No. WH48. 10 pp.

Loh, I.C. and Stokes, R.A. (1981). Predicting stream salinity changes in south-western Australia. *Agric. Water Manage.* **4:** 227–54.

Loveday, J. and Bridge, B.J. (1983). Management of salt-affected soils. *In* 'Soils an Australian viewpoint'. CSIRO, Australia. 843–56.

Lumis, G.P., Hofstra, G. and Hall, R. (1973). Sensitivity of roadside trees and shrubs to aerial drift of deicing salts. *HortScience* **8:** 556–61.

Maas, E.V. and Hoffman, G.J. (1977). Crop salt tolerance–current assessment. *J. Irrig. Drainage Div. Am. Soc. Civil. Eng.* **103:** 115–34.

MacLaren, J.P. (1995). Radiata pine growers manual. New Zealand Forest Research Iinstitute Bulletin 184. 140 pp.

Macpherson, D.K. and Peck, A.J. (1987). Models of the effect of clearing on salt and water export from a small catchment. *J. Hydrol.* **94:** 163–80.

Malcolm, C.V. and Allen, R.J. (1981). The Mallen niche seeder for plant establishment on difficult sites. *Aust. Rangelands J.* **3:** 106–9.

Malik, M.N. and Sheikh, M.I. (1983). Planting of trees in saline and waterlogged areas. I. Test planting at Azakhel. *Pakistan J. For.* **33:** 1–17.

Marcar, N.E. (1989). Salt tolerance of frost resistant eucalypts. *New Forests* **3:** 141–9.

Marcar, N.E. (1990). Tree option for utilization of salt-affected land. *In* (Eds. Myers, B.A. and West, D.W.) Revegetation of Saline Land. Proceedings of a Workshop held at the Institute for Irrigation and Salinity Research, 29–31 May, Tatura, Victoria: Institute for Irrigation and Salinity Research. 53–9.

Marcar, N.E. (1992). Trees for salt-affected land. *In* National Workshop on Productive Use of Saline Land. Waite Agricultural Research Institute, Department of Agriculture, Adelaide. p. 12–23.

Marcar, N.E. (1993). Waterlogging modifies growth, water use and ion concentrations in seedlings of salt-*treated Eucalyptus camaldulensis, E. tereticornis, E. robusta* and *E. globulus*. *Aust. J. Plant Physiol.* **20:** 1–13.

Marcar, N.E. (1995). Eucalypts for salt-affected and acid soils. *In* (Eds. K.G. Eldridge, M.P. Crowe and K.M. Old) Environmental management: the role of eucalypts and other fast growing species. Proc. Joint Australian/Japanese Workshop, Canberra. 90–9.

Marcar, N.E. and Termaat, A. (1990). Effects of root-zone solutes on *Eucalyptus camaldulensis* and *Eucalyptus bicostata* seedlings responses to Na^+, Mg^{++} and Cl^-. *Plant and Soil* **28:** 1–10.

Marcar, N.E., Crawford, D.F. and Leppert, P.M. (1991a). The potential of trees for utilisation and management of salt-affected land. *In* (Eds N. Davidson, and R. Galloway). Productive use of saline land. Proceedings of a workshop held in Perth, Western Australia, 10–14 May. ACIAR Proceedings No 42 17–22.

Marcar, N.E., Hussain, R.W., Haq, I., Ansari, R. and Arunin, S. (1991b). Preliminary results of field trials with Australian native tree species on salt-affected land in Pakistan and Thailand. p. 85. *In* (Ed. P.J. Ryan) Proc. Third Aust Forest Soils and Nutrition Conference, Productivity in Perspective. Melbourne 7–11 October, Forestry Commission of NSW, Sydney. 252 pp.

Marcar, N.E., Crawford, D.F., Leppert, P.M. and Nicholson, A. (1994). Tree growth on saline land in southern Australia. *In* (Eds. M.S. Schulz and G. Petterson). Productive Use of Saline Lands. Saline Irrigation Areas. Proceedings of a Workshop, Echuca, Australia. 15–17 March. 60–5.

Marcar, N.E., Crawford, D., Leppert, P., Jovanovic, T., Floyd, R. and Farrow, R. (1995). 'Trees for Saltland. A Guide to Selecting Native Species for Australia'. CSIRO Division of Forestry, Melbourne. 72 pp.

Marcar, N.E. and Khanna, P.K. (1997). Reforestation of salt-affected and acid soils. *In* (Eds E.K.S. Nambiar and A.G. Brown) Management of soil, nutrients and water in tropical plantation forests. ACIAR Monograph No. 43. 481–526.

Martin, J.P., Richards, S..J. and Pratt, P.F. (1964). Relationship of exchangeable Na percentage at different soil pH levels to hydraulic conductivity. *Soil Sci. Soc. Amer. Proc.* **28:** 620–2.

Mathur, N.K. and Sharma, A.K. (1984). *Eucalyptus* in reclamation of saline and alkaline soils in India. *Indian Forester* **110:** 9–15.

McLennan, W. (1996). Australians and the Environment. Australian Bureau of Statistics, Commonwealth of Australia, Canberra. 415 pp.

McComb, J.A., Bennett, I.J., van der Moezel, P.G. and Bell, D.T. (1989). Biotechnology enhances utilization of Australian woody species for pulp, fuel and land rehabilitation. *Aust. J. Biotechnol.* **3:** 297–301.

McGrath, D., Ward, D., Jenkins, P.J. and Read, B. (1991). Influence of site factors on the productivity and drought susceptibility of *Pinus radiata* in the Blackwood Valley region of Western Australia. *In* (Ed. P.J. Ryan) Proc. Third Aust Forest Soils and Nutrition Conference, Productivity in Perspective. Melbourne 7–11 October, Forestry Commission of New South Wales, Sydney. 65–6.

McNeill, J.D. (1980). Electromagnetic terrain conductivity measurement at low induction numbers. Technical Note TN-6: Geonics Pty Ltd, Ontario, Canada.

Midgely, S.J., Eldridge, K.G. and Doran, J.C. (1989). Genetic resources of *Eucalyptus camaldulensis*. *Comm. For. Rev.* **68:** 295–308.

Milthorpe, P.L., Hillan, J.M. and Nicol, H.I. (1994). The effect of time of harvest, fertiliser and irrigation on dry matter and oil production of blue mallee. *Industrial Crops and Products* **3:** 165–73.

Morabito, D., Mills, D., Prat, D. and Dizengremel, P. (1994). Response of clones of *Eucalyptus microthera* to NaCl *in vitro*. *Tree Physiology* **14:** 201–10.

Morris, J. D. (1981). Factors affecting the salt tolerance of eucalypts. *In* (Eds K.M. Old, G.A. Kile and C.P. Ohmart). Eucalypt Dieback in Forests and Woodlands. CSIRO Publications, Melbourne. Proceedings of a Conference held at CSIRO Division of Forest Research, Canberra, 4–6 August 1980. 190–204.

Morris, J.D. (1984). Comparative salt-tolerance of ornamental trees and shrubs for Victorian field planting. Forests Commission Victoria, Research Branch Report 249. 20 pp. (Unpublished).

Morris, J.D. (1995). Clonal red gums for Victorian planting. Trees and Natural Resources, March 1995. 26–8.

Morris, J.D. and Thomson, L.A.J. (1983). The role of trees in dryland salinity control. *Proc. Royal Soc. Victoria* **95:** 123–31.

Morris, J.D. and Wehner, B.A. (1987). Daily and annual wateruse by four eucalypt species irrigated with waste water at Robinvale. Report 329. Department of Conservation, Forests and Lands, Victoria. 30 pp.

Morris, J.D., Bickford, R. and Collopy, J. (1994). Tree and shrub performance and soil conditions in a plantation irrigated with saline groundwater. Department Conservation and Natural Resources, Forest Research and Development Branch Research Paper 357. 37 pp.

Morrow, P.A., Bellas, T.E. and Eisner, T. (1976). *Eucalyptus* oils in the defensive oral discharge of Australian sawfly larvae (Hymenoptera: Pergidae). *Oecologia* **24:** 193–206.

Morrow, P.A. and Fox, L.A. (1980). Effect of variation in *Eucalyptus* essential oil yield on insect growth and grazing damage. *Oecologia* **45:** 209–19.

Munns, R. and Termaat, A. (1986). Whole plant responses to salinity. *Aust. J. Plant Physiol.* **13:** 143–60.

Murray–Darling Basin Ministerial Council (1989). Draft Murray–Darling Basin natural resources management strategy. Canberra: Murray–Darling Basin Ministerial Council. 19 pp.

Myers, B.J., Bond, W.J., Benyon, R.G., Falkiner, R.A., Polglase, P.J., Smith, C.J., Snow, V.O and Theiveyanathan, S. (1999). Sustainable Effluent-irrrigated Plantations: An Australian Guideline. CSIRO Forestry and Forest Products. Canberra. 293 pp.

Myers, B.J., Theiveyanathan, S., O'Brien, N.D. and Bond, W.J. (1996). Growth and water use of *Eucalyptus grandis* and *Pinus radiata* plantations irrigated with effluent. *Tree Physiology* **16:** 211–9.

Naidu, R., Rengasamy, P., de Lacy, N.J. and Zarcinas, B.A. (1995). Soil solution composition of some sodic soils. (Eds R. Naidu, M.E. Sumner and P. Rengasamy) 'Australian sodic soils: Distribution, properties and management'. CSIRO Publications, Melbourne. 155–70.

Norman, C.P. (1989). Kyvalley EM-38 salinity survey, August 1988. Research Report Series No. 82, January 1989, Institute for Sustainable Agriculture, Department of Agriculture and Rural Affairs, Victoria.

Norman, C.P. (1990). Monitoring soil salinity using electromagnetic induction (EM-38) in the Campaspe Irrigation Area, Victoria. Institute for Sustainable Agriculture, Department of Agriculture and Rural Affairs, Victoria.

Norman, C.P. and Heath, J. (1992). Soil salinity survey of the Girgarre irrigated woodlot trial site. May 1992. Tatura Salinity Research Program. Institute for Sustainable Agriculture, Department of Agriculture and Rural Affairs, Tatura Centre, Victoria. 25 pp.

Northcote, K.H. and Skene, J.K.M. (1972). Australian Soils with Saline and Sodic Properties. CSIRO Australia, Soil Pub. No 27.

Nulsen, R.A. and Baxter, I.N. (1982). The potential of agronomic manipulation for controlling salinity in Western Australia. *J. Aust. Inst. Agric. Sci.* **48:** 222–6.

Oddie, R.L.A., McComb, J.A. and Bell, D.T. (1996). *Eucalyptus camaldulensis* x *globulus* hybrids for saline land. p. 367-371. *In* Proc. 4th National Conference and Workshop on the Productive Use and Rehabilitation of Saline Lands, Albany, Western Australia. 25-30 March 1996. Promaco Conventions Pty Ltd, Canning Bridge, Western Australia.

Olbrich, B.W., Le Roux, D., Poulter, A.G., Bond, W.J. and Stock, W.D. (1993). Variation in water use efficiency and $\delta^{13}C$ levels in *Eucalyptus grandis* clones. *J. Hydrol.* **150:** 615–34.

Oldeman, L.R., van Engelen, V.W.P. and Pulles, J.H.M. (1991). The extent of human-induced soil degradation. *In* Olderman, L.R., Hakkeling, R.T.A. and Sombroek, W.G. World Map of the Status of Human-Induced Soil Degradation: An Explanatory Note. Waggeningen: International Soil Reference and Information Centre (ISRIC). 27–33.

Parsons, R.F. (1968). Effects of waterlogging and salinity on growth and distribution of three mallee species of *Eucalyptus*. *Aust. J. Bot.* **16:** 101–8.

Pearce, S. (1999). Natural resource management policy statement. Its development and significance. Outlook 99, Canberra. 107–40.

Pearman, G.I. (1989). Greenhouse gases: evidence for atmospheric changes and anthropogenic causes. *In* (Ed. G.I. Pearman) 'Greenhouse: Planning for climate change'. CSIRO, Melbourne. 3–21.

Peck, A.J. (1977). Development and reclamation of secondary salinity. *In* (Eds J.S. Russell and E.L. Greacen) 'Soil Factors in Crop Production in a Semi-Arid Environment'. University of Queensland Press, St Lucia, Qld.

Peck, A.J. (1978). Salinisation of non-irrigated soils and associated streams: A review. *Aust. J. Soil Res.* **16:** 157–68.

Peck, A.J. (1983). Response of groundwaters to clearing in Western Australia. *In* Papers of the International Conference on Groundwater and Man, Australian Water Resources Council Conference, Australian Government Public Service, Canberra. Series No. 8, Volume 2. 327–35.

Peck, A.J. and Hurle, D.H. (1973). Chloride balance of some farmed and forested catchments in southwestern Australian. *J. Water Resources Res.* **9:** 648–57.

Peck, A.J. and Williamson, D.R. (1987). Effects of forest clearing on groundwater. *J. Hydrology* **94:** 47–66.

Pepper, R.G. and Craig, G.F. (1986). Resistance of selected *Eucalyptus* species to soil salinity in Western Australia. *J. Appl. Ecol.* **23:** 977–87.

Pettit, N.E. and Froend, R.H. (1992). Research into reforestation techniques for saline groundwater control. Report No. WS97. Water Authority of Western Australia, Leederville, Western Australia. 42 pp.

Pierce, L.L., Walker, J., Dowling, T.I., McVicar, T.R., Hatton, T.J., Running, S.W. and Coughlan, J.C. (1993). Ecohydrological changes in the Murray Darling Basin. III. A simulation of regional hydrological changes. *J. Appl. Ecol.* **30:** 283–94.

Polglase, P.J., Myers, B.J., Theiveyanathan, S., Falkiner, R.A., Smith, C.J. (1994). Growth, water and nutrient balances in the Wagga Effluent Plantation Project: Preliminary Results. Report to Expanding Horizons–Recycled Water Seminar. NSW Recycled Water Co-ordination Committee and Australian Water and WasteWater Association. Newcastle May 1994. 9 pp.

Poulter, D.G. and Chaffer, L.C. (1991). Dryland salinity: some economic issues. *Agriculture and Resources Quality* 3(3): 361-370.

Prat, D. and Riyad, A.F. (1990). Variation in organic and mineral components in young *Eucalyptus* seedlings under saline stress. *Physiologia Plantarum* **79:** 479–86.

Prenzel, J. (1979). Mass flow to the root systems and mineral uptake of a beech stand calculated from 3-year field data. *Plant and Soil* **51:** 39–49.

Prinsley, R.T. (1992). The role of trees in sustainable agriculture – an overview. *Agroforestry Systems* **20:** 87–115.

Pryor, L.D. and Johnson, L.A.S. (1971). A classification of the eucalypts. Australian National University, Canberra.

Ranasinghe, D.M.S. and Mayhead, G.J. (1991). Dry matter content and its distribution in an age series of *Eucalyptus camaldulensis* (Dehn.) plantations in Sri Lanka. *For. Ecol. Manage.* **41:** 137–42.

Raper, G.P. (1998). Agroforestry water use in Mediterranean Regions of Australia. RIRDC Publication No. 98/63. 71 pp.

Rayment, G.E. and Higginson, F.R. (1992). 'Australian laboratory handbook of soil and water chemical methods. Organic carbon'. Inkata Press, Melbourne, Australia. 29–30.

Rhoades, J.D. and Corwin, D.L. (1981). Determining soil electrical conductivity versus saturation paste–depth relations using an inductive electromagnetic soil conductivity meter. *Soil Sci. Soc. Am. J.* **45:** 255–60.

Rhoades, J.D., Manteghi, N.A., Shouse, P.J. and Alves, W.J. (1989). Soil electrical conductivity and soil salinity: New formulations and calibrations. *Soil Sci.* **53:** 433–9.

Rhoades, J.D. and Miyamoto, S. (1990). Testing soils for salinity and sodicity. *In* (Ed. R.L. Westerman) 'Soil Testing and Plant Analysis', Third Edition. Soil Science Society of America Book Series No. 3, Soil Science Society of America, Madison, Wisconsin. 299– 336.

Rhodes, D. and Felker, P. (1988). Mass screening of *Prosopis* (mesquite) seedlings for growth at seawater salinity concentrations. *For. Ecol. Manage.* **24:** 169–76.

Richards, L.A. (1969). Diagnosis and improvement of saline and alkali soils. USDA Agric. Handbook No. 60.

Robertson, G.A. (1993). Salinity in Australia – A national perspective. In Land Management for Dryland Salinity Control. Proc. National Conference. 9–12.

Robertson, G.A. (1996). Saline lands in Australia – its extent and predicted trends*In* Proc. 4th National Conference and Workshop on the Productive Use and Rehabilitation of Saline Lands, Albany, Western Australia. 25–30 March. Promaco Conventions Pty Ltd, Canning Bridge, Western Australia. 43–7.

Rogers, A.L. (1985). Foliar salt in *Eucalyptus* species. *Aust. For. Res.* **15:** 9–16.

Ruprecht, J.K. and Schofield, N.J. (1989). Analysis of streamflow generation following deforestation in south-west Western Australia. *J. Hydrol.* **105:** 1–17.

Ruprecht, J.K. and Schofield, N.J. (1991a). Effects of partial deforestation on hydrology and salinity in high salt storage landscapes. I: Extensive block clearing. *J. Hydrology* **129:** 19–38.

Ruprecht, J.K. and Schofield, N.J. (1991b). Effects of partial deforestation on hydrology and salinity in high salt storage landscapes. II: Strip, soil and parkland clearing. *J. Hydrology* **129:** 39–55.

Ryan, P.A. (1993a). Hardwood plantations in Queensland? Why not! Notes on a Seminar presented at Queensland Forest Research Institute, Gympie, Queensland, 13 August 1993. Queensland Department of Primary Industries Forest Service.

Ryan, P.A. (1993b). Evaluation and development of tropical and subtropical Australian woody species in Queensland. Proc. Institute of Foresters of Australia 15th Biennial Conference 'Australasian Forestry and the Global Environment', Queensland, 19–24 September. 15 pp.

Salama, R.B., Bartle, G.A. and Farrington, P. (1994). Water use of plantation *Eucalyptus camaldulensis* estimated by groundwater hydrograph separation techniques and heat pulse method. *J. Hydrol.* **156:** 163–80.

Salinity Audit of the Murray–Darling Basin (1999). A 100-year perspective. Murray–Darling Basin Commission Office, Canberra, ACT.

Sands, R. (1981). Salt resistance in *Eucalyptus camaldulensis* Dehn. from three different seed sources. *Aust. For. Res.* **11:** 93–100.

Sands, R. and Clarke, A.R.P. (1977). Response of radiata pine to salt stress. I. Water relations, osmotic adjustment and salt uptake. *Aust. J. Plant Physiol.* **4:** 637–46.

Sardabi, H. (1991). A trial of exotic species in the Eastern Caspian Littoral of Iran. p. 168. *In* (Ed. P.J. Ryan) Proc. Third Aust Forest Soils and Nutrition Conference, Productivity in Perspective. Melbourne 7–11 October, Forestry Commission of NSW, Sydney. 168.

Savory, B.M. (1962). Boron deficiency in eucalypts in northern Rhodesia. *Emp. For. Rev.* **41:** 118–26.

Schofield, N.J. (1989). Trees in salinity control. *In* (Ed. P.T. Arkell) Trees – Everybody Profits. Australian Institute of Agricultural Science, Occasional Publication 45. 25–40.

Schofield, N.J. (1991). Effects of tree plantations on groundwater. *Agric. Sci.* July 1991: 26–9.

Schofield, N.J. (1992). Tree planting for dryland salinity control in Australia. *Agroforestry Systems* **20:** 1–23.

Schofield, N.J., Loh, I.C., Scott, P.R., Bartle, J.R., Ritson, P., Bell, R.W., Borg, H., Anson, B. and Moore, R. (1989). Vegetation strategies to reduce stream salinities of water resource catchments in south-west Western Australia. Report to the Steering Committee for Research on Land Use and Water Supply. Water Authority of Western Australia, Report No WS 33. 98 pp.

Schofield, N.J. and Bari, M.A. (1991). Valley reforestation to lower saline groundwater tables: results from Stene's Farm, Western Australia. *Aust. J. Soil Res.* **29:** 635–50.

Schofield, N.J. and Scott, P. (1991). Planting trees to control salinity. W.A. J. Afric. **32:** 3–10.

Schrale, G., Boardman, R. and Blaskett, M.J. (1993). Investigated land based disposal of Bolivar reclaimed water, South Australia. *Water Science and Technology* **27:** 87–96.

Schreuder, G.F., de Barros, A.A.A. and Hill, D.A. (1990). The global supply and cost-price structure of *Eucalyptus*. *In* (Ed. D.F. Root) Global resources and markets: issues and trends. 2nd International Symposium on Pulp and Paper. College of Forest Resources, University of Washington, Seattle, Washington, USA. Inst. For. Res. Contribution No. 71. 81–92.

Sena Gomes, A.R. and Kozlowski, T.T. (1980). Effects of flooding on *Eucalyptus camaldulensis* and *Eucalyptus globulus* seedlings. *Oecologia (Berl.)* **46:** 139–42.

Shannon, M.C. (1985). Principles and strategies in breeding for higher salt tolerance. *Plant and Soil* **89:** 227–41.

Shannon, M.C., Suhayda, C.G., Grieve, C.M., Grattan, S.R., Francois, L.E., Poss, J.A., Donovan, T.J., Draper, J.H. and Oster, J.D. (1997). Water use of *Eucalyptus camaldulensis*, Clone 4544, in saline drainage reuse systems. *In* Proc. California Plant and Soil Conference. California Chapter American Society Agronomy & California Fertilizer Association. 15 & 16, February. 20–8.

Shannon, M.C., Grieve, C.M., Wilson, C., Poss, J.A. and Guzy, M.R. (1998). Growth and water relations of plant species suitable for saline drainage water reuse systems. Final Report California Department of Water Resources, Project DWR B-59922. 89 pp.

Sharma, M.L., Barron, R.J.W. and Williamson, D.R. (1987). Soil water dynamics of lateritic catchments as affected by forest clearing for pasture. *J. Hydrol.* **94:** 29–46.

Shaw, R. (1999). Soil salinity–electrical conductivity and chloride. *In* (Eds K.I. Peverill. L.A. Sparrow and D.J. Reuter) 'Soil Analysis: An Interpretation Manual'. CSIRO Publishing, Melbourne, Australia. 129-45.

Shaw, S., Boardman, R. and Schrale, G. (1996a). Effluent irrigated plantations: An alternative to ocean disposal. *In* Proc. Symposium on Irrigation Australia 96, Adelaide. Department of Primary Industries, South Australia. 9 pp.

Shaw, S., Boardman, R. and Schrale, G. (1996b). The prospects for effluent irrigated plantations: Biomass production and water use of five native hardwood species, 23rd Hydrology and Water Resources Symposium, Hobart, Australia. 21–24 May.

Shaybany, B. and Kashirad, A. (1978). Effect of NaCl on growth and mineral composition of *Acacia saligna* in sand culture. *J. Am. Soc. Hortic. Sci.* **103:** 823–6.

Sheikh, M.I. (1974). Afforestation in waterlogged and saline areas. *Pakistan J. Forestry* **24:** 186–92.

Shiva, M.P., Prakash, O.M., Singh, S. and Jaffer, R. (1988). Role of eucalypts in agroforestry and essential oil production potential. *Indian Forester* **114:** 776–83.

Singh, B. (1989). Rehabilitation of alkaline wasteland on the Gangetic alluvial plains of Upper Pradish, India, through afforestation. *Land Degradation and Rehabilitation* **1:** 305–10.

Slavich, P.G. (1990). Determining EC_a depth profiles from electromagnetic induction measurements. *Aust. J. Soil Res.* **28:** 443–52.

Slavich, P.G. and Petterson, G.H. (1990). Estimating average root-zone salinity from electromagnetic induction (EM-38) measurements. *Aust. J. Soil Res.* **28:** 453–63.

Slavich, P.G. and Petterson, G.H. (1993). Estimating the electrical conductivity of saturated paste extracts from 1:5 soil:water suspensions and texture. *Aust. J. Soil. Res.* **31:** 73–81

Smith, S.T. (1962). Some aspects of soil salinity in Western Australia. Faculty of Agriculture, MSc Thesis, University of Western Australia.

Sokal, R.R. and Rohlf, F.J. (1981). Biometry. W.H. Freeman and Co. San Francisco USA. 859 pp.

Sonogan, R.M.C. and Patto, R.M. (1985). The use of trees and other vegetation in the management of seepage from earthen channels in the Wimmera-Mallee region. Progress Report. Rural Water Commission of Victoria.

Stackpole, D., Borschmann, R. and Baker, T. (1995). Commercial tree-growing for land and water care. 3. Establishment of pilot sites. A Report to the Trees for Profit Research Board, Produced by the Centre for Forest Tree Technology and The University of Melbourne, Australia. 57 pp.

Steering Committee of the Climate Change Study (SCCCS) (1995). Climate change science: current understandings and uncertainties. Australian Academy of Technological Sciences and Engineering. 100 pp.

Stewart, H.T.L., Flinn, D.W., Baldwin, P.J. and James, J.M. (1981). Diagnosis and correction of iron deficiency in planted eucalypts in north-west Victoria. *Aust. For. Res.* **11:** 185–90.

Stewart, H.T.L. and Flinn, D.W. (1984). Establishment and early growth of trees irrigated with wastewater at four sites in Victoria, Australia. *For. Ecol. Manage.* **8:** 243–56.

Stewart, H.T.L., Hopmans, P., Flinn, D.W., Hillman, T.J. and Collopy, J. (1988). Evaluation of irrigated tree crops for land disposal of municipal effluent at Wodonga. Albury-Wodonga Development Corporation Technical Paper. No. 7. 31 pp.

Stewart, H.T.L., Hopmans, P., Flinn, D.W. and Croatto, G. (1990). Harvesting effects on phosphorus availability in a mixed eucalypt ecosystem in southeastern Australia. *For. Ecol Manage.* **36:** 149–62.

Stolte, W.J., McFarlane, D.J. and George, R.J. (1997). Flow systems, tree plantations and salinisation in a Western Australian catchment. *Aust. J. Soil Res.* **35:** 1213–29.

Stone, C. (1993). Fertilizer and insecticide effects on tree growth and psyllid infestation of young *Eucalyptus grandis* and *E. dunnii* plantations in northern New South Wales. *Aust. For.* **56:** 257–63.

Stone, C. and Bacon, P.E. (1994). Relationships among moisture stress, insect herbivory, foliar cineole content and the growth of river red gum *Eucalyptus camaldulensis*. *J. Appl. Ecol.* **31:** 604–12.

Stribbe, E. (1975). Soil moisture depletion in summer by a *Eucalyptus* grove in a desert area. *Agro-Ecosystem* **2:** 117.

Sun, D. and Dickinson, G.R.D. (1993). Responses to salt stress of 16 *Eucalyptus* species, *Grevillea robusta, Lophostemon confertus* and *Pinus caribaea* var. *hondurensis*. *For. Ecol. Manage.* **60:** 1–14.

Sun, D. and Dickinson, G.R.D. (1995a). Survival and growth responses of a number of Australian tree species planted on a saline site in tropical north Australia. *J. Appl. Ecol.* **32:** 817–26.

Sun, D. and Dickinson, G.R.D. (1995b). Salinity effects on tree growth, root distribution and transpiration of *Casuarina cunninghamiana* and *Eucalyptus camaldulensis* planted on a saline soil in tropical north Australia. *For. Ecol. Manage.* **77:** 127–38.

Sykes, S.R. (1985). A glasshouse screening procedure for identifying citrus hybrids which restrict chloride accumulation in shoot tissues. *Aust. J. Agric. Res.* **36:** 779–89.

Szabolcs, I. (1981). Salt-affected soils. Boca Raton, Florida. CRC Press, 274 pp.

Tal, M. (1985). Genetics of salt tolerance in higher plants: theoretical and practical considerations. *Plant and Soil* **89:** 199–226.

Tanji, K.K. and Karajeh, F.F. (1993). Saline drain water reuse in agroforestry systems. *J. Irrig. Drainage Eng.* **119:** 170–80.

Thomson, L.A.J. (1988). Salt tolerance in *Eucalyptus camaldulensis* and related species. Ph D Thesis. University of Melbourne.

Thomson, L.A.J., Morris, J.D. and Halloran, G.M. (1987). Salt tolerance in eucalypts. *In* (Ed. R.S. Rana) Afforestation of Salt-affected Soils. Proc. Internat. Symp., Karnal, India. Volume 3. 1–12.

Tomar, O.S. and Gupta, R.K. (1985). Performance of some forest tree species in saline soils under shallow and saline water-table conditions. *Plant and Soil* **87:** 329–35.

Tome, M. and Periera, J.S. (1991). Growth and management of eucalypt plantations in Portugal. *In* (Ed. P.J. Ryan) Proc. Third Aust Forest Soils and Nutrition Conference, Productivity in Perspective, Melbourne 7–11 October, Forestry Commission of New South Wales, Sydney. 147–57.

Totey, N.G., Kulkarni, R., Bhowmik, A.K., Khatri, P.K., Dahia, J.K. and Prasad, A. (1987). Afforestation of salt affected wasteland. I. Screening for tree species of Madhye Prodest for salt tolerance. *Indian Forester* **113:** 805–15.

Townsend, A.M. (1980). Response of selected tree species to sodium chloride. *J. Amer. Soc. Hort. Sci.* **105:** 878–83.

Townsend, A.M. (1989). The search for salt tolerant trees. *Arboricultural J.* **13:** 67–73.

Truman, R. and Turner, J. (1972). Mineral deficiency symptoms in *Eucalyptus pilularis*. Forestry Commission of NSW, Technical Paper **18:** 21 pp.

Truman, R. and Lambert, M.J. (1978). Salinity damage to Norfolk Island Pines caused by surfactants. I. The nature of the problem and effect of potassium, sodium and chloride concentration on uptake by roots. *Aust. J. Plant Physiol.* **5:** 377–85.

Turner, J. (1972). An initial survey of foliar chloride levels in introduced conifers in New South Wales. *In* (Ed. R. Boardman) The Australian Forest-Tree Nutrition Conference, 1971. Forestry and Timber Bureau, Canberra. 168–82.

Turner, J. (1982). The mass flow component of nutrient supply in three western Washington forest types. *Oecologia Plantarum* **3:** 323–9.

Turner, J. and Kelly, J. (1973). Foliar chloride levels in some eastern Australian plantation forests. *Soil Sci. Soc. Am. Proc.* **37:** 443–5.

Turner, J. and Lambert, M.J. (1986). Effects of forest harvesting nutrient removals on soil nutrient reserves. *Oecologia (Berlin)* **70:** 140–8.

Turner, J. and Lambert, M.J. (1988). Soil properties as affected by *Pinus radiata* plantations. *N.Z. J. For. Sci.* **18:** 77–91.

Turner, J., Thompson, C.H., Turvey, N.D., Hopmans, P. and Ryan, P.J. (1990). A soil technical classification system for *Pinus radiata* (D. Don) plantations. I. Development. *Aust. J. Soil Res.* **28:** 797–811.

Turner, J. and Lambert, M.J. (1991). Nutrient cycling processes in *Eucalyptus grandis* forests on the New South Wales north coast. *In* (Ed. P.J. Ryan) Proc. Third Aust Forest Soils and Nutrition Conference, Productivity in Perspective. Melbourne 7–11 October, Forestry Commission of New South Wales, Sydney. 198–9.

Turner, J., Lambert, M.J. and Holmes, G.I. (1992). Nutrient cycling in forested catchments in south-eastern New South Wales. 1. Biomass accumulation. *For. Ecol. Manage.* **55:** 135–48.

Turner, J., Lambert, M.J. and Knott, J. (1996a). Nutrient inputs from rainfall in New South Wales State Forests. State Forests of New South Wales Research Paper No 17. 49 pp.

Turner, J., Knott, J. and Lambert, M.J. (1996b). Fertilization of *Pinus radiata* plantations after thinning. I. Productivity gains. *Aust. For.* **59:** 7–21.

Turner, J., Lauck, V., Dawson, J. and Lambert, M.J. (1996c). Water quality monitoring strategies for forest management: A case study at Bago State Forest. State Forests of New South Wales Research Paper No 30. 25 pp.

Turner, J., Lambert, M.J. and Dawson, J. (1996d). Water quality monitoring strategies for forest management: A case study in the Towamba Valley Catchment. State Forests of New South Wales Research Paper No 31. 32 pp.

Turner, J. and Lambert, M.J. (1999). Change in organic carbon in forest plantation soils in eastern Australia. *For. Ecol. Manage.* (in press).

Turvey, N.D. Ed. (1987). A technical classification for soils of *Pinus* plantations in Australia: field manual. School of Forestry, The University of Melbourne. Bulletin No. 6. 42 pp.

van der Moezel, P.G. and Bell, D.T. (1987). Comparative seedling salt tolerance of several *Eucalyptus* and *Melaleuca* species from Western Australia. *Aust. For. Res.* **17:** 151–8.

van der Moezel, P.G., Watson, L.E., Pearce-Pinto, G.V.N. and Bell, D.T. (1988). The response of six *Eucalyptus* species and *Casuarina obesa* to the combined effects of salinity and waterlogging. *Aust. J. Plant Physiol.* **15**: 465–74.

van der Moezel, P.G., Walton, C.S., Pearce-Pinto, G.V.N. and Bell, D.T. (1989a). Screening for salinity and water-logging tolerance in five *Casuarina* species. *Landscape and Urban Planning* 17: 331-337.

van der Moezel, P.G., Watson, L.E. and Bell, D.T. (1989b). Gas exchange responses of two *Eucalyptus* species to salinity and waterlogging. *Tree Physiology* **5**: 251–7.

van der Moezel, P.G. and Bell, D.T. (1990). Saltland reclamation: selection of superior Australian tree genotypes for discharge sites. *Proc. Ecol. Soc. Aust.* **16**: 545–9.

van der Moezel, P.G., Pearce-Pinto, G.V.N. and Bell, D.T. (1991). Screening for salt and water-logging tolerance in *Eucalyptus* and *Melaleuca* species. *For. Ecol. Manage.* **40**: 27–37.

Vertessy, R.A., Dawes, W.R., Zhang, L., Hatton, T.J. and Walker, J. (1995). Catchment scale hydrologic modelling to assess the water and salt balance behaviour of eucalypt plantations. *In* (Eds K.G. Eldridge, M.P. Crowe and K.M. Old). 'Environmental management: the role of eucalypts and other fast growing species'. CSIRO, Melbourne. 110–21.

Wagner, R. (1989). Dryland salinity in the South East Region. Soil Conservation Service of NSW, Sydney.

Walker, G.R., Budd, G.R., Pavelic, P., Kennett-Smith, A.K. and Cook, P.G. (1990). Groundwater recharge beneath open woodlands in south-western New South Wales. Centre for Groundwater Processes. Report No. 22. 32 pp.

Walker, R.R., Blackmore, D.H. and Qing, S. (1993). Carbon dioxide assimilation and foliar ion concentrations in leaves of lemon (*Citrus limon* L.) trees irrigated with NaCl or $NaSO_4$. *Aust. J. Plant Physiol.* **20**: 173–85.

Ward, S.C. (1991). Processes of land degradation and the role of plantations in its amelioration. *In* (Ed. P.J. Ryan) Proc. Third Aust Forest Soils and Nutrition Conference, Productivity in Perspective. Melbourne 7–11 October, Forestry Commission of NSW, Sydney. 71–9.

Watson, R.T., Meira Filho, L.G., Sanhueza, E. and Janetos, A. (1992). Greenhouse gases: sources and sinks. *In* (Eds J.T. Houghton, B.A. Callander, and S.K. Varney). 'Climate Change 1992: The supplementary Report to the IPCC Scientific Assessment', Cambridge University Press, Cambridge. 29–46.

Weston, C.J. (1991). Factors limiting the growth of eucalypts across a range of sites in Gippsland, Victoria. *In* (Ed. P.J. Ryan) Proc. Third Aust Forest Soils and Nutrition Conference, Productivity in Perspective. Melbourne 7–11 October 1991, Forestry Commission of NSW, Sydney. 158–9.

Will, G.M. (1961a). Some notes on nutrient deficiency in *Eucalyptus* spp. Second World *Eucalyptus* Conference, Sao Paulo, Brazil. Reports and Documents, II: 938–41.

Will, G.M. (1961b). Some changes in the growth habits of eucalypt seedlings caused by nutrient deficiencies. *Emp. For. Res.* **40**: 301–7.

Williams, P.J. (1979). Reforestation as an economic solution to salinity control. *In* Water Resources and Land Management Issues in the Darling Range. Institute of Engineers, Australian Hydrology Symposium, Perth, Western Australia, August. Section 8.1.

Williamson, D.R. (1978). Water is the villain in dryland salinity. *Victoria's Resources* **20**: 9–14.

Williamson, D.R. (1990). Salinity – an old environmental problem. Year Book Australia 1990, 202–10.

Williamson, D.R., Stokes, R.A. and Ruprecht, J.K. (1987). Response of input and output of water and chloride to clearing for agriculture. *J. Hydrol.* **94**: 1–29.

Williamson, D.R., Gates, G.W.B., Robinson, G., Linke, G.K., Seker, M.P. and Evans, W.R. (1997). Salt Trends. Historic trend in salt concentration and saltload of stream flow in the Murray–Darling Drainage Division. Murray–Darling Basin Commission, Dryland Technical Report No 1. 64 pp.

Winter, K., Osmond, C.B. and Pate, J.S. (1981). Coping with salinity. *In* (Eds J.S. Pate and A.J. McComb) 'The Biology of Australian Plants', University of Western Australia Press, Nedlands, Western Australia. 88–113.

Woods, R.V. (1955). Salt deaths in *Pinus radiata* at Mt Crawford Forest Reserve, S.A. *Aust. For.* **19**: 13–9.

Yadav, J.S.P. (1980). Salt affected soils and their afforestation. *Indian Forester* **106**: 259–72.

Yadav, J.S.P. (1989). Problems and potentials of reforestation of salt-affected soils in Sri Lanka. Report of Regional W/E Development Program in Asia GCP/RAS/I I I/NET. FAO Field Document 16, Bangkok. 17 pp.

Yadav, J.S.P. and Singh, K. (1970). Tolerance of certain forest species to varying degrees of salinity and alkali. *Indian Forester* **96:** 587–99.

Yanosky, T.M., Hupp, C.R. and Hackney, C.T. (1995). Chloride concentrations in growth rings of *Taxodium distichum* in a saltwater-intruded estuary. *Ecol. Appl.* **5:** 785–92.

Yanosky, T.M. and Kappel, W.M. (1997). Effects of solution mining of salt on wetland hydrology as inferred from tree rings. *Water Resources Research* **33:** 457–70.

Yeo, A.R. (1994). Physiological criteria in screening and breeding. *In* (Eds A.R. Yeo and T.J. Flowers) Monographs on Theoretical and Applied Genetics. Vol 21. Springer-Verlag, Berlin. 37–59.

Young, A. (1987). The environmental basis of agroforestry in meteorology and agroforestry. (Eds W.E. Reilsnyder and T.O. Darnholer) ICRAF, Nairobi. 29–48.

Zohar, Y. (1979). Plantations of eucalypt on drained lake peat in the Hula Valley. *La-Yaaran* **29:** 43–9.

Zohar, Y. (1982). Growth of eucalypts on saline soils in the Wadi'Arava. *In* Contribution from Agricultural Research Organization, Bet Dagan, Israel No. 1266-H. 1982 Series. 60–4.

Zohar, Y. and Schiller, G. (1998). Growth and water use by selected seed sources of *Eucalyptus* under high water table and saline conditions. *Agriculture, Ecosystems and Environment* **69:** 265–77.

Glossary

Acid soil The chemical activity of hydrogen ions in soil expressed in terms of pH.

Aggregration The formation of aggregates or granites by a variety of processes within the soil.

Agroforestry The integration of forest trees within an agricultural system with the objective of obtaining benefit from both.

Alkaline soil The chemical condition of soil with a pH greater than 7.0. Often associated with saline soils.

Anaerobic Describes soil conditions in which free oxygen is deficient and in which chemically reducing processes prevail. Such conditions are usually found in waterlogged or poorly drained soil in which water has replaced soil air.

Aquifer A porous soil or geological formation, often lying between impermeable subsurface strata, which holds water and through which water can percolate slowly over long distances and which yields ground water to springs and wells. Aquifers may, however, be unconfined and the water level subject to seasonal inflow.

Available nutrient The portion of any element or compound in the soil that can be taken up and is required by plants to enhance their growth and development.

Available soil water That part of the water in the soil that can be absorbed by plant roots. The amount of water held between the moisture content prevailing at any point in time and the moisture content at which plant growth ceases.

Biomass The mass of plant and animal material usually expressed as mass per unit area (g/m^2 or kg/ha), excluding organic matter in the soil.

Buffer capacity The ability of a soil to resist changes in pH. The buffering action is due mainly to the properties of clay and fine organic matter. Thus, with the same pH level, more lime is required to neutralise a clayey soil than a sandy soil, or a soil rich in organic matter than one low in organic matter.

Bulk density The mass of dry soil per unit bulk volume. The bulk volume is determined before drying to constant weight at 105°C. The unit of measurement is usually grams per cubic centimetre.

Calcareous Refers to materials, particularly soils, containing significant amounts of calcium carbonate. Describes rocks composed largely of, or cemented by, calcium carbonate.

A calcareous soil is one containing carbonate in sufficient quantity to effervesce visibly when treated with cold dilute (N) hydrochloric acid.

Carbon accretion The accumulation of carbon by plants, usually forest tree species. Expressed as mass per unit area or where discussing the rate of accretion, as mass/unit area/unit time.

Cation exchange capacity The total amount of exchangeable cations that a soil can adsorb, expressed in centimoles of positive charge per kilogram of soil. Cations are positive ions such as calcium, magnesium, potassium, sodium, hydrogen, aluminum and manganese, these being the most important ones found in soils. Cation exchange is the process whereby these ions interchange between the soil solution and the clay or organic matter complexes in the soil. The process is very important as it has a major controlling effect on soil properties and behaviour, stability of soil structure, the nutrients available for plant growth, soil pH and the soil's reaction to fertilisers and other ameliorants added to the soil.

Chlorosis Condition in plants relating to the failure of chlorophyll to develop. Chlorotic leaves range from light green to yellow to almost white.

Clay Soil material consisting of mineral particles less than 0.002 mm in equivalent diameter. This primarily includes the chemically active mineral part of the soil. Many of the important physical and chemical properties of a soil depend on the type and quantity of clay it contains. Three broad classes of clay type are recognised, namely, montmorillonite, kaolinite and illite.

Clearing Removal of existing woody vegetation from an area, usually to convert the area to a different land use.

Colloidal material The finest clay and organic material with a particle size generally less than 10^{-4} mm in diameter. Such material represents the finest particles removed in an erosion event, and as such remains permanently in suspension, unless subject to flocculation.

Current annual increment Refers to the growth of forests over a specific twelve month period.

Cyclic salt Salt deposited on soils from wind or rainfall. Near the sea or inland salt lakes, where the amounts deposited are likely to be significant, such salt may subsequently be leached into the soil and take part in various chemical processes there.

Dispersible soil A structurally unstable soil which readily disperses into its constituent particles (clay, silt, sand) in water. Highly dispersible soils are normally highly erodible and are likely to give problems related to field and earthwork tunnelling.

Duplex soil A soil in which there is a sharp change in soil texture between the A and B horizons (such as loam overlying clay). The soil profile is dominated by the mineral fraction with a texture contrast between the A and B horizons of 1.5 soil texture groups or greater. Horizon boundaries are clear to sharp. The texture change from the bottom of the A horizon to the top of the B horizon occurs over a vertical distance of 10 cm or less.

Effective soil depth The depth of soil material that plant roots can penetrate readily to obtain water and plant nutrients. It is the depth to a layer that differs sufficiently from the overlying material in physical or chemical properties to prevent or severely retard the growth of roots.

Effluent Contaminated water derived from agricultural, industrial or domestic sources.

Electrical conductivity A measure of the conduction of electricity through water or a water extract of soil. It can be used to determine the soluble salts in the extract and hence soil salinity. The unit of electrical conductivity is the Siemens and soil salinity is normally expressed as milliSiemens per centimetre at 25°C.

Evaporation Loss of water to the atmosphere from the soil surface or water body.

Evapotranspiration The combined loss of water from a given area, and during a specified period of time, by evaporation from the soil surface and by transpiration from plants.

Exchange capacity The total ionic charge of a soil, expressed in centimoles of charge per kilogram of soil Its numerical value is identical to the value expressed in milliequivalents per 100 grams of soil.

Exchangeable sodium percentage (ESP) The extent to which the adsorption complex of a soil is occupied by sodium. It is expressed as follows: ESP = exchangeable sodium (centimoles/kg soil)/Cation exchange capacity (centimoles/kg soil) x 100.

Field capacity The percentage of water remaining in a soil after free drainage has practically ceased, usually two to three days after having been saturated.

Fertiliser Any organic or inorganic material of natural or synthetic origin which is added to a soil to supply certain elements essential to the growth of plants.

Gradational profile A soil in which there is a gradual change in soil texture between the A and B horizons (for example, loam over clay loam over light clay). The soil is dominated by the mineral fraction and shows more clayey texture grades on passing down the solum of such an order that the texture of each successive horizon changes gradually to that of the one below. Horizon boundaries are usually gradual or diffuse.

Ground water Subsurface water contained in a saturated zone of soil or geological strata and capable of moving in response to gravity and hydraulic pressure gradients.

Halophyte Plant that is adapted to very salty soil.

Horizon A layer of soil, approximately parallel to the surface, with distinct characteristics produced by soil-forming processes.

Hydraulic conductivity An expression of the readiness with which a liquid such as water flows through a soil in response to a given potential gradient.

Hydrological cycle The continuous interchange of water between land, sea or other water surface and the atmosphere.

Humus That more or less stable fraction of the soil organic mater remaining after the major portion of added plant and animal residues have been decomposed. Usually it is dark coloured.

Infiltration The downward movement of water into the soil. It is largely governed by the structural condition of the soil, the nature of the soil surface including presence of vegetation, and the antecedent moisture content of the soil.

Interception The proportion of incident precipitation captured by foliage or crowns and evaporated from their surface.

Irrigation Artificial addition of water to an area to overcome a deficiency affecting crop or plant growth.

Leachate Liquid containing dissolved minerals and salts as a result of leaching. Under natural conditions, it refers to seepage water containing dissolved minerals and salts.

Leaf area index The ratio of the total surface area of a plant's leaves to the ground area of the plant.

Leaching The removal in solution of the more soluble minerals and salts by water seeping through soil, rock, ore body or waste material.

Litter The uppermost layer of organic material in a soil, consisting of freshly fallen or slightly decomposed organic materials which have accumulated at the ground surface.

Lysimeter A device to measure the quantity or rate of water movement through or from a block of soil, usually undisturbed and *in situ*, or to collect such percolated water for quality analysis.

Mean annual increment Forest growth averaged over a time period.

Merchantable volume Quantity of saleable wood produced by a forest, usually expressed in cubic metres.

Mottling The presence of more than one soil colour in the same horizon, not including different nodule colours. The subdominant colours normally occur as scattered blobs or blotches, which have definable differences in hue, value or chroma from the dominant colour. Mottling is often indicative of slow internal drainage, but may also be a result of parent material weathering.

Nutrient cycling The uptake, utilisation and return of nutrients to the soil by plant communities. Geochemical cycling refers to the inputs and losses from the system while biological cycling refers to the cycling within the plant and soil system. The nutrient pools are in units of mass/unit area (for example, g/m^2) while transfers such as litterfall and uptake are in mass/unit area/unit time (for example, $g/m^2/yr$).

Organic soil A soil in which soil organic matter dominates the profile. The surface 30 cm should contain 20% or more organic matter if the clay content of the mineral soil is 15% or lower, or 30% or more organic matter if the clay content of the mineral soil is higher than 15%.

Osmotic pressure A type of pressure exerted in living bodies as a result of unequal concentration of salts on both sides of a cell wall or membrane. Water will move from the part having the least salt concentration through the membrane into the part having the highest salt concentration and, therefore, exerts additional pressure on this side of the membrane.

Parent material The geologic material from which a soil profile develops. It may be bedrock or unconsolidated materials including alluvium, colluvium, aeolian deposits or other sediments.

pH A measure of the acidity or alkalinity of a soil. A pH of 7.0 denotes neutrality, higher values indicate alkalinity and lower values indicate acidity. Strictly, it represents the negative logarithm of the hydrogen ion concentration in a specified soil/water suspension on a scale of 1 to 14. Soil pH levels generally fall between 5.5 and 8.0 with most plants growing best in this range.

Piezometer A narrow tube, open at each end, inserted down a hole in the ground to the depth of the water table, which enables measurement of its elevation or hydraulic head.

Plantation Artificial establishment of forest stands by planting.

Precipitation The deposition of water in a solid or liquid form on the earth's surface from atmospheric sources.

Provenance The geographical source, place or origin of a given lot of seeds or plants.

Pulpwood Wood that is used to make pulp, usually for further processing into paper.

Recharge The addition of water to a aquifer from all sources.

Saline soil A soil which contains sufficient soluble salts to adversely affect plant growth and/or land use. Generally, a level of electrical conductivity of a saturation extract in excess of 4 mS/cm at 25°C.

Sawlog A log removed from a forest suitable for conversion to sawn timber.

Site specific management The manipulation of a crop or forest according to defined characteristics of the area to meet a specific objective. The manipulation includes selection of suitable species.

Sodic soil A soil containing sufficient exchangeable sodium to adversely effect soil stability, plant growth and/or land use. Such soils would typically contain a horizon in which the ESP or amount of exchangeable sodium expressed as a percentage of cation exchange capacity would be six or more. The soils would be dispersible and may be improved by the addition of gypsum. Strongly sodic soils are considered to be those with an ESP of 15% or more.

Sodium adsorption ratio (SAR) SAR = sodium/(square root of the sum of calcium and magnesium divided by 2). The cation concentrations are in milliequivalents per litre.

Soil aeration The process by which air in the soil is replenished by air from the atmosphere. In a well-aerated soil, the soil air is similar in composition to the atmosphere above the soil. Poorly aerated soils usually contain a much higher percentage of carbon dioxide and a correspondingly lower percentage of oxygen. The rate of aeration depends largely on the volume and continuity of pores in the soil.

Soil classification The systematic arrangement of soils into groups or categories on the basis of similarities and differences in their characteristics. Soils can be grouped according to their genesis (taxonomic classification), their morphology (morphological classification), their suitability for different uses (interpretative classification) or according to specific properties.

Soil degradation Decline in soil quality commonly caused through its improper use by humans. Soil degradation includes physical, chemical and/or biological deterioration. Examples are loss of organic matter, decline in soil fertility, decline in structural condition, erosion, adverse changes in salinity, acidity or alkalinity and the effects of toxic chemicals, pollutants or excessive flooding.

Soil fertility The capacity of the soil to provide adequate supplies of nutrients in proper balance for the growth of specified plants, when other growth factors such as light, moisture and temperature are favourable. The more general concept of soil fertility can be divided into three components: chemical etc, physical etc, biological etc.

Soil horizon A layer of soil material within the soil profile with distinct characteristics and properties which are produced by soil-forming processes, and which are different from those of the layers below and/or above. Generally horizons are more or less parallel to the land surface, except that tongues of material from one horizon may penetrate neighbouring horizons.

Soil organic matter That fraction of the soil including plant and animal residues at various stages of decomposition, cells and tissues of soil organisms, and substances synthesised by them. It is of vital importance as it contributes to the cation exchange and field capacities of the soil and provides a major source of plant nutrients and substances which assist in soil structure maintenance.

Soil permeability The characteristic of a soil, soil horizon or soil material which governs the rate at which water moves through it. It is a composite expression of soil properties and depends largely on soil texture, soil structure, the presence of compacted or dense soil horizons and the size and distribution of pores in the soil. The rate varies widely in the field, from more than 3000 mm/day in poorly graded sands and gravels, to less than 0.01 mm/day for some heavy plastic clays.

Soil profile A vertical cross-sectional exposure of a soil, extending downwards from the soil surface to the parent material or, for practical purposes, to a depth of 1 m where the parent material cannot be differentiated. It is generally composed of three major layers designated A, B and C horizons. The A and B horizons are layers that have been modified by weathering and soil development and comprise the solum. The C horizon is weathering parent material which has not, as yet, been significantly altered by biological soil-forming processes.

Soil salinity The characteristic of soils relating to their content of water-soluble salts. Such salts predominantly involve sodium chloride, but sulphates, carbonates and magnesium salts occur in some soils. High salinity adversely affects the growth of plants, and therefore increases erosion hazard. Soil salinity is normally characterised by measuring the electrical conductivity of a soil/water saturation extract and is expressed in milliSiemens per centimetre at 25°C.

Soil solution The water in a soil containing ions dissociated from the surfaces of soil particles, and other soluble substances.

Soil structure The combination or spatial arrangement of primary soil particles (clay, silt, sand, gravel) into aggregates such as peds or clods and their stability to deformation. Structure may be described in terms of the grade, class and form of other soil aggregates.

Soil texture The coarseness or fineness of soil material as it affects the behaviour of a moist ball of soil when pressed between the thumb and forefinger. It is generally related to the proportion of soil particles of differing sizes (sand, silt, clay and gravel) in a soil, but is also influenced by organic matter content, clay type and degree of structural development of the soil.

Soil type A soil having a morphology which is distinct from the surrounding soils, but comprises such a limited geographic area that the delineation of a new map unit, or the naming of a new soil taxonomic unit, is not justified.

Soil water Water contained in, or in transit by drainage through, the soil.

Solution That part of natural weathering processes whereby substances in rocks and soil are dissolved in ground water. Such water usually contains some carbon dioxide and organic compounds from plant breakdown which make it mildly acid, thereby enhancing its ability to dissolve rock materials.

Symptoms Changes in appearance of plant parts related to deficiency or toxicity of one or more chemical elements.

Total dissolved solids (TDS) A term that expresses the quantity of dissolved material in a sample of water, either the residue on evaporation dried at 105°C or for many waters that contain more than about 1000 mg/L, the sum of the chemical constituents.

Toxicity The characteristic of a soil relating to its content of elements or minerals which adversely affect plant growth. It is of particular concern in relation to acid soils. Soils with pH less than 5.0 may give rise to manganese and aluminium toxicities, which reduce plant growth and hence ground cover. It is also of concern in the rehabilitation of heavy metal mines, where toxic levels of such elements as copper, zinc and lead in mine tailings create difficulties in their revegetation.

Uniform soil A soil in which there is little, if any, change in soil texture between the A and B horizons (for example, loam over loam, sandy clay over silty clay). The soil is dominated by the mineral fraction and shows minimal texture differences throughout, such that no clearly defined texture boundaries are to be found. The range of texture throughout the solum is not more than the equivalent span of one soil texture group.

Veneer Thin section of timber produced by peeling or slicing of logs and used as a facing product for plywood or for laminated veneer lumber.

Water balance An estimated state of equilibrium within the soil moisture regime based on rainfall, evapotranspiration, run-off, drainage and soil moisture storage. Soil moisture is assumed to severely limit plant growth in months when rainfall, together with an antecedent moisture content up to a maximum of 100 mm, is less than 40% of pan evaporation. At other times, soil moisture is regarded as being adequate for plant growth. When the assumed field capacity of the soil is reached, any further rainfall is regarded as being lost as run-off.

Watertable The upper surface of ground water or that level below which the soil is saturated with water; locus of points in soil water at which the hydraulic pressure is equal to atmospheric pressure.

Waterlogged The condition of a soil which is saturated with water and in which most or all of the soil air has been replaced. The condition which is detrimental to most plant growth, may be caused by excessive rainfall, irrigation or seepage, and is exacerbated by inadequate site and/or internal drainage.

Water-repellant soils (Hydrophobic soils) Soils which resist wetting when dry. Drops of water do not spread spontaneously over their surface and into pores. The degree of water repellence may be severe where water drops remain on a flattened surface for some minutes. In other cases, drops appear to be absorbed readily but quantitative measurements show that the height of capillary rise is diminished.

INDEX